**Bent Rolf Pettersen**

# The Atom – An Organism

**Copyright © 2024 by Nova Science Publishers, Inc.**
https://doi.org/10.52305/MQCW3008

**All rights reserved.** No part of this book may be reproduced, stored in a retrieval system or transmitted in any form or by any means: electronic, electrostatic, magnetic, tape, mechanical photocopying, recording or otherwise without the written permission of the Publisher.

We have partnered with Copyright Clearance Center to make it easy for you to obtain permissions to reuse content from this publication. Please visit copyright.com and search by Title, ISBN, or ISSN.

For further questions about using the service on copyright.com, please contact:

Copyright Clearance Center
Phone: +1-(978) 750-8400 Fax: +1-(978) 750-4470 E-mail: info@copyright.com

## NOTICE TO THE READER

The Publisher has taken reasonable care in the preparation of this book but makes no expressed or implied warranty of any kind and assumes no responsibility for any errors or omissions. No liability is assumed for incidental or consequential damages in connection with or arising out of information contained in this book. The Publisher shall not be liable for any special, consequential, or exemplary damages resulting, in whole or in part, from the readers' use of, or reliance upon, this material. Any parts of this book based on government reports are so indicated and copyright is claimed for those parts to the extent applicable to compilations of such works.

Independent verification should be sought for any data, advice or recommendations contained in this book. In addition, no responsibility is assumed by the Publisher for any injury and/or damage to persons or property arising from any methods, products, instructions, ideas or otherwise contained in this publication.

This publication is designed to provide accurate and authoritative information with regards to the subject matter covered herein. It is sold with the clear understanding that the Publisher is not engaged in rendering legal or any other professional services. If legal or any other expert assistance is required, the services of a competent person should be sought. FROM A DECLARATION OF PARTICIPANTS JOINTLY ADOPTED BY A COMMITTEE OF THE AMERICAN BAR ASSOCIATION AND A COMMITTEE OF PUBLISHERS.

## Library of Congress Cataloging-in-Publication Data

ISBN: 979-8-89113-734-9 (Softcover)
ISBN: 979-8-89113-789-9 (eBook)

Published by Nova Science Publishers, Inc. † New York

# Contents

# List of Figures

# Chapter 1

# Introduction

Atoms are energies concentrated, stored and organized in a system. Different energies are stored in different quarks. The quarks are put together in atomic elements like protons and neutrons. These elements are organized into atoms.

Different atoms have different quark compositions and number of atomic elements. This gives atoms' their properties and our periodic system.

Atoms performs different tasks including, among others:

- *An attractive force*. This creates attraction and form atomic bonds and gravity.
- *A repulsive force*. If atomic cores collide, they can be destroyed. Atoms therefore produce a repulsive force which prevents the nuclei from colliding and help stabilize and position the atoms.
- *The strong nuclear force*. The strong nuclear force help organize quarks and atomic elements into a fixed system. The strong force keeps the atomic elements together.
- *Communication tasks*. We know through quantum entanglement atoms and particles can communicate over great distances. Atoms also seem to communicate with atoms close by and therefore also appear to be aware of their proximity.

Atoms seem to be able to adjust these actions. It looks like they adjust their tasks based upon their situation and preferences. Their main task is to preserve their structure. They appear to be able to adjust actions based on information received. This call for a new evaluation of the nature of atoms. If atoms seem to act on information, they might be looked upon in an organic metaphor.

In this book I will go into and attempt to explain these actions. It will also explain gravity as a force working between matter and universal phenomena. The conversion of energies in atoms, Atomic Phase Displacement, will be explained.

Atoms continuously use a lot of energy to perform these tasks. To be in energy balance, the atoms therefore must simultaneously receive the same amount of energy in return. Energy balance will be explained.

The creation of particles and matter will be explained. Also how atoms might evolve into dark matter.

All this is based on a new model, the ilefos model. The ilefos model explains all these elements. Matter, dark matter, dark energy and gravity can be explained in the model. The model works in all fields. It works in quantum mechanics and astrophysics, thus bridging between nuclear physics and astrophysics.

Are atoms “intelligent”? Do they have an operating system and intelligent design? This book will discuss this and suggests answers.

Atomic bonds give an indication of other bonds in the universe. If you can understand the bonds in an atom, you can understand the universe.

Try to understand the universe and different dimensions and you will understand everything.

Understand the atom and you will understand the universe.

# Chapter 2

# Energies

An energy can create a force. We know several forces in atoms:

- The attractive force pulls atoms toward one another. It creates an attraction between atoms and makes atomic bindings and gravity.
- The strong force that concentrates energies in systems (quarks) and form atomic elements (protons/neutrons), and hold these together in an atomic system.
- The repulsive force. The attractive force pulls atoms towards each other's. To prevent collisions between the atom' nuclei, which can destroy the atomic core, the atoms have a repulsive force. This force helps the atoms to keep their distance and helps keep them in a "fixed" position.

These are some examples of forces in atoms. Each force must be caused by a special energy.

We also know atoms communicates with each other. We see this in quantum entanglement where particles/atoms can communicate over large distances and synchronize their spin.

Atoms seem to share energies with each other. They seem to prefer dielectricity as energy for sharing. When atoms receive more energy than they can store, atoms appear to prefer dielectricity to get rid of excess energy. But they also may use different channels.

Atoms also seem to be able to convert energies to the energies the atoms need. We see this in an electro magnet where atoms convert electricity to magnetism. If we reverse the process, atoms can change magnetism to electricity in a generator.

Atoms seem to continuous handle several energies.

# Chapter 3

# Quarks

Energies are concentrated in different quarks. Different quarks store and handles different energies. These quarks are organized in structures by the strong nuclear force.

Different atoms can have different number of quarks in their atomic elements (protons/neutrons). The quark composition of the atomic elements gives atoms its properties. This tells us how much of specific energies an atom can handle, thus what it prefers.

A quark contains immense energy. We can see examples of this when atoms release energy through nuclear fission and atomic dissolution. In stars and nuclear explosions we see how quarks can release their energy.

Later we will take a closer look at different types of quarks and control quarks.

# Chapter 4

# Atoms

Atoms are quarks put into a system. When quarks are put under central management, they can be formed into atomic elements. Protons and neutrons are atomic elements.

The actions of an atom are controlled by a control quark, the atomic quark. This quark controls all atomic elements and all quarks in an atom.

Through the strong nuclear force, the atomic quark is able to hold the quarks in a solid structure. This structure is an atom.

An atom consists of:

- Proton (s)
- Neutron (s)
- Uniparticle (s)

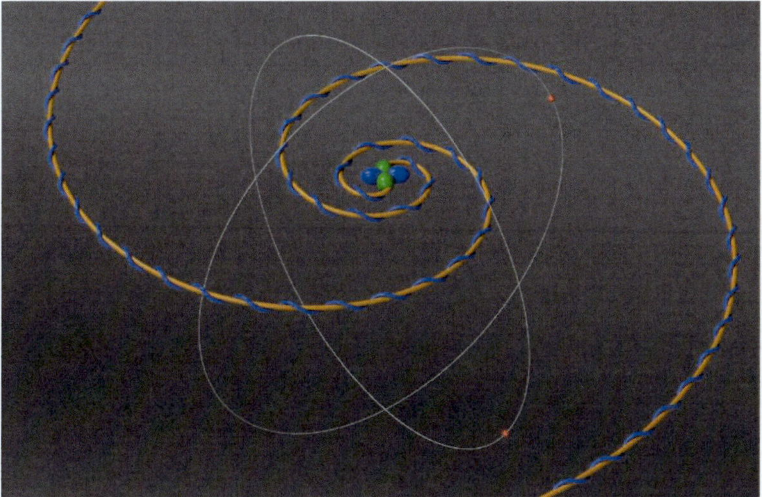

**Figure 1.** An atom with protons, neutrons, ilefos- and dielectricity track and uni particles (Stian Kongsvik).

All atoms have at least one of each element. Except hydrogen, which do not have a neutron. Here the proton has to handle the tasks usually performed by the neutron.

The uniparticle is a particle which circles around the atomic core. It performs special tasks, which we will come back to in a later chapter.

# Chapter 5

# Atomic Actions

The atom continuously performs several tasks including, among others:

- The strong force
- The attractive force
- The repulsive force
- Dielectricity
- Communication tasks
- Management tasks

The atom's operating system manages these tasks through the control quark, the atomic quark.

## The Strong Force

**Figure 2.** The strong nuclear force working between two quarks (Ola Tandstad).

The strong force attracts and controls all quarks. All quarks consist of quark material. This material is attracted to strong force. The atomic quark (control quark) administrates this energy. Through the strong force the atomic quark is able to put different quarks into atomic elements. An atomic element can be a proton or a neutron.

These elements are held together by the strong force in a structure, which forms the atom. The strong force is therefore the most important force in the atom.

## The Attractive Force

The attractive force attracts other atoms. This force is sent out from neutrons in attractive force tracks. The number of neutrons determines the number of attractive tracks from the atom.

Hydrogen does not have a neutron. In hydrogen the proton has to make the attractive force. The attractive force of hydrogen is therefore weak compared to other atoms.

Different types of atoms have different quark compositions. The quark composition determines which energies the atom prefers and also the strength of the attractive force.

The attractive force is caused by special energy units, ilefos energy units. These are released from the neutron. They attract each other. Due to the attractive force between the units, the units form a track of ilefos energy units when they are released from the neutron. This is a gravity track which I call ilefos track.

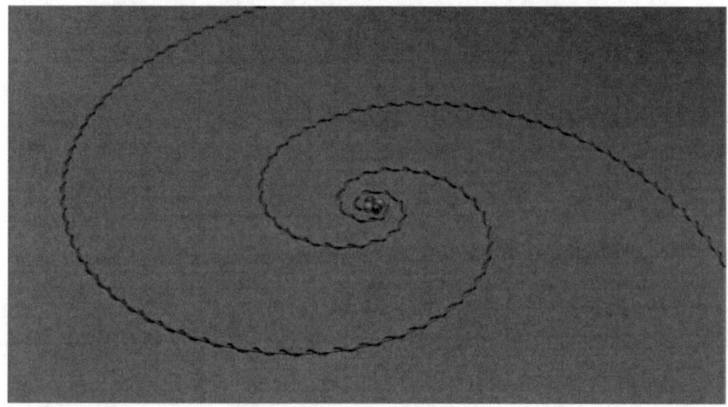

**Figure 3.** An atom with ilefos track (Stian Kongsvik).

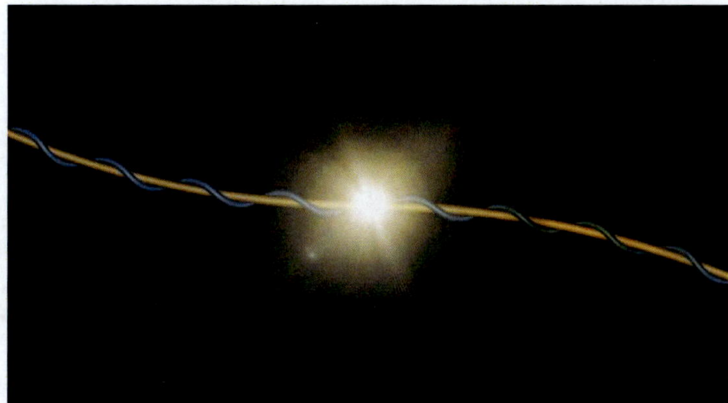

**Figure 4.** Two ilefos tracks meet and form a binding (Stian Kongsvik).

The ilefos track (gravity track) have attraction to ilefos tracks from other atoms. This creates a pull (attraction) between atoms.

The ilefos track (gravitational track) is strong close to the source, then its strength gradually reduces as distance increases.

When two ilefos tracks meets, they form an ilefos bond. If the track is strong, the two atoms form a strong connection, an atomic bond.

If two weaker tracks meet, they form a weaker bond, gravity.

The attractive force pulls atoms towards each other.

## The Repulsive Force

The repulsive force helps atoms keep distance to each other. Without the repulsive force, the atoms' nuclei (atomic cores) would crush each other due to the attractive force. The attractive force pulls atoms toward each other, the repulsive force helps atoms keep a distance. The balance between the attractive- and repulsive force help atom position each other in fixed positions.

The repulsive force in an atom varies. The strength of the attractive force determines the repulsive force. In addition, the repulsive force may even increase when atomic cores are in danger of collision.

If two atoms are moving fast towards each other, the atom will increase the repulsive force to prevent a collision between the nuclei.

The repulsive force is caused by uni energy through uni particles. Around each atom we have special particles with repulsive force. These particles circle the atom in fixed orbits. Uni particles will be explained in a special chapter later in the book.

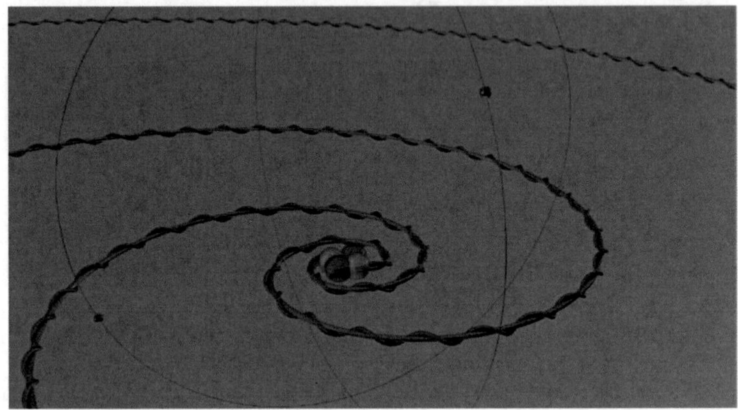

**Figure 5.** Uni particles in orbit around the nucleus (Stian Kongsvik).

## Dielectricity

Atoms share energy to be assured of an even energy level among the atoms. They seem to communicate with each other and are aware of the state of neighbour atoms. Energy seems to be evenly shared between the atoms.

We see this when heat when heat distributes in matter. Electricity is concentrated energy, which is distributed though atoms. Different types of atoms have different energy transport capacities based on their quark composition. Atoms with more energy quarks in their atomic elements, are able to store and transport larger amounts of energy. Copper is an example of such an element.

Energy is transported between atoms in energy tracks. An energy track is energy units which due to mutual attraction, form tracks of energy units. Dielectricity tracks are tracks formed by dielectricity energy units.

Since energy seem to be transported in separated tracks and not by particles (electrons), I refer to the energy as dielectricity to separate the term from electricity.

If houses are to share energy, they do not run around and exchange batteries (electrons). They build electronic wires between the houses (energy tracks). Atoms also use the most effective way to share energy, through energy tracks.

Atoms convert energies to perform their tasks. Dielectricity is the universal energy which is easiest to share and convert. This is therefore the preferred method of sharing energy between atoms.

## Communication Tasks

Through quantum entanglement we know atoms can communicate over large distances. Particles or matter (atoms), can be synchronized. They then have synchronized spin. If they are separated, they keep the same spin. If we change the spin of one particle, the other also changes its spin. This happens instantly. Instantly mean the particles communicate and synchronize their spin with a speed much faster than the speed of light. This effect can be observed over several hundred kilometres.

What does this tell us? Atoms can be synchronized and communicate. This communication is much faster than the speed of light. The speed of light is not the speed limit. Dimensional energies, which are not affected by gravity, are not limited by the speed of light.

We know atoms can communicates over larger distances through quantum entanglement. We therefore must assume atoms also use different energies to communicate over short distances when they "talk" to each other. This communication makes atoms able to communicate and be aware of their surroundings. They can then share energies and act upon changes in their surroundings. When atom cores, nuclei, are in danger of colliding, atoms increase their repulsive force to prevent the destruction of the nuclei. We have several other examples, which will be described later in the book.

## Management Task

Atoms are able to adjust their actions based on feedback from their surroundings. They are able to convert energies based on their needs.

To be able to act upon feedback and perform tasks, atoms must have an operating system. They must have an atomic quark, a management and control quark. This quark controls all function of an atom. The atomic quark controls all atomic elements and particles in the atom.

The atom also needs special quarks who carry the information needed for controlling the atom. I refer to these as special quarks. They are decentralized quarks which give instructions to the atomic quark. They are program quarks which handle subfunctions in the atom.

Together with the special quarks the atomic quark is able to organize and manage the atom.

The atomic quark uses a specific energy to perform its tasks.

## Atomic DNA

Atoms seem to have an "atomic DNA" which govern each type of atom's structure. This gives each type of atom in the periodic system its properties. The type and number of specific quarks in an atomic element, and the number of atomic elements (protons/ neutrons), give the specific atom its properties.

"Atomic DNA" might be an incorrect term. DNA means deoxyribonucleic acid which is linked to the genes. It carries the genetic instructions for development in living organisms. But in the absence of an existing term, I use "Atomic DNA." The atomic DNA give instructions for function and development of an atom.

If an atom is exposed to immense energy, it can grow. We see this in supernova explosions and other places with very strong energy, where atoms can grow and fusion. The atomic DNA give instructions for how atoms grow. This will be explained in a separate chapter.

To sum up, these are the main actions of an atom:

- The strong force
- The attractive force
- The repulsive force
- Dielectricity
- Communication tasks
- Management tasks

We must assume atoms also perform other tasks, but these seem to be the main tasks.

Based on these actions and properties, the atom must be a complex and advanced system. It has an atomic DNA, and converts and handles several different energies. It seems to be able to be aware of its surroundings. Later we will give examples how it can act on received information to preserve its structure.

All these properties suggest the atom is an advanced structure which may behave like an organism. Therefore the atom might be looked upon in an organismic metaphor.

# Chapter 6

# Proton

An atomic core consists of atomic elements, proton (s) and neutron (s). The exception is hydrogen which generally only has a proton.

Proton s and neutrons have the same nuclear quark composition, but they have different behaviour. Why?

**Figure 6.** A proton (Stian Kongsvik).

The properties of a nuclear element are determined by the quark composition and how the quarks are managed. Each atomic element, for example a proton, have a local management inside each element. This local management is performed by a quark called the negative quark.

The negative quark controls the proton. The quark has a quark which works together with it, a special quark. This quark is a program quark which hold all the programs and instructions. The negative quark holds a decentralized operating system, the special quark holds the different application programs.

The negative quarks in the protons are identical in composition and operating system. They tell which energies and actions the proton shall prioritize and perform, by instructions from the atomic quark.

The neutrons also have negative quarks which govern the neutrons' actions and which energies the neutron shall prioritize. The neutron negative quarks have slightly different operating system, which make the neutron behave in a different way compared to protons.

The proton focusses on energy. The main energy in a proton is dielectricity. This energy is stored, managed and converted in the energy quark(s).

A proton always has at least one energy quark. The number of energy quarks determine the energy handling capacity of the proton.

A proton with only one energy quark has moderate energy handling capacity.

A proton with two or more energy quarks has higher energy handling capacity. Oxygen and copper are examples of atoms with high energy capacity. These are able to store, release and transfer much energy. (The difference between gas and solid matter will be explained later).

Matters with large electronic resistance have protons with only one energy quark.

Protons prioritize dielectricity and are responsible for converting energies to dielectricity. Atoms with several energy quarks in their protons can convert much energy to dielectricity. In a generator we see how copper, atoms with several energy quarks in the protons, convert pulses of magnetism to dielectricity.

Protons also have other tasks in the atom. They also handle dimensional and communication tasks.

# Chapter 7

# Neutron

Each neutron also has a decentralized operating system which govern the element. The negative quark in the neutron is responsible for controlling and managing the neutron. The neutron's negative quark has a slightly different operating system compared to the protons. The neutrons have the same quark composition, but the negative quark prioritize different actions and energies.

The neutrons prioritize the energies of forces, especially the attractive and repulsive force.

**Figure 7.** A neutron (Stian Kongsvik).

The attractive force is energy units which attract each other. They pull similar energy units toward themselves. The plus quark is responsible for the repulsive force. I call this energy ilefos and the energy units ilefos energy units. This quark might be the Higgs boson.

When a neutron releases ilefos energy units (attractive energy units), the mutual attraction of the units makes the units form a track of ilefos energy units when they are sent out from the neutron, an ilefos track. Each neutron sends out an ilefos track (gravity track). These ilefos tracks attract external ilefos tracks from other atoms. When two ilefos tracks meet, we have a bounding between the atoms. The strength of the track determines the strength the atomic binding creates.

The ilefos track is strong when released from the neutron. The track loose energy with distance from origin, thus the track becomes weaker farther from the atom.

When two different ilefos tracks connects close to the atom, we have a strong binding, an atomic binding. When ilefos tracks from two different atoms meet further away, the tracks are weaker. We then have a weaker attraction between the atoms, gravity.

The plus quark handles the ilefos energy in the neutron. The number of plus quarks in the neutron determine the ilefos capacity of the neutron. Atoms with only one plus quark in their neutrons have weak ilefos capacity and thus weaker ilefos tracks. Gases are atoms with only one plus quark in their neutrons. A weak ilefos track creates a weaker atomic bound with external atoms.

Metals are examples of atoms with three or more plus quarks in their protons. This gives the atoms greater ilefos capacity, thus stronger ilefos tracks. The stronger ilefos tracks creates stronger atomic bonds, as we see in metals.

Neutrons also handles the repulsive force. The TE quark handles the repulsive energy. I call the repulsive energy Uni energy.

The uni energy is repulsive towards ilefos. It is therefore transported to special particles which handle the uni energy. I call these particles uni particles. Each neutron manages a uni particle. See next chapter for more information on the repulsive force and the uni particles.

Both the neutrons and the protons produce and handle the strong force. The neutrons also handle dimensional energies and other tasks.

# Chapter 8

# Uni Particles

TE quarks in atomic elements produce uni power, the repulsive force. An atomic core cannot release both attractive and repulsive force. They will neutralize each other.

Therefore the neutrons send out ilefos tracks, gravity tracks with attractive force, directly from the neutron.

How do we best protect the nucleus from collisions? The nucleus needs a protective shield. This shield also has to allow ilefos tracks, the attractive force, to go out from the nucleus.

This is fixed by the atom by having special particles with repulsive force in orbit around the nucleus. These particles are called uni particles. The uni particles have repulsive force towards other uni particles and ilefos tracks.

Each atom has uni particles in orbit. These will repel uniparticles from other atoms, thus prevent nuclei collisions and help position atoms.

Uni particles has repulsive force towards ilefos tracks (gravity tracks) from neutrons. This help stabilize the ilefos track and prevent them from spinning.

An atomic element (proton/neutron) has 256 or more quarks. More information on this will come later. An uniparticle has 10-15 quarks. The size is determined by the size of nucleus. A "heavy" atom needs stronger uni particles than a lighter nucleus. A heavy atom has stronger ilefos tracks and attractive force and therefore need stronger repulsive force (uni particles).

Each neutron controls an uniparticle. The number of uni particles are therefore equal to the number of neutrons in the nucleus. The exception is hydrogen. Hydrogen has no neutron. Here the tasks of the neutron are performed by the proton. This means that the proton also controls uni particle and ilefos track. Therefore, the uni particle and ilefos tracks of hydrogen is very weak.

The uni particle is almost an atomic element. It has a negative quark controlling its functions, TE quark producing uni energy, dimensional quarks which allow communication with the neutron and other uni particles. But it is controlled by the neutron, not by the atomic quark. It has a decentralized operating system and is not directly controlled by atomic quark which manages protons and neutrons. Therefore I characterize uni particles as particles rather than atomic elements.

When an atom senses a risk of collision, the atomic quark orders the neutrons to produce more repulsive force. The neutrons then give instructions to the uni particles to produce more repulsive force. Neutrons may also send more energy to the particles if needed.

The uni particles go in orbit at a great distance from the nucleus. In the illustrations I have placed them close to the nucleus to give a clear picture of actions, but they are further away than the illustrations should suggest. If the nucleus was 1 cm (0.4 inch), the uni particles will orbit around 800 meters (875 yards) from the nucleus.

The main parts of an atom are: proton(s), neutron(s) and uniparticles.

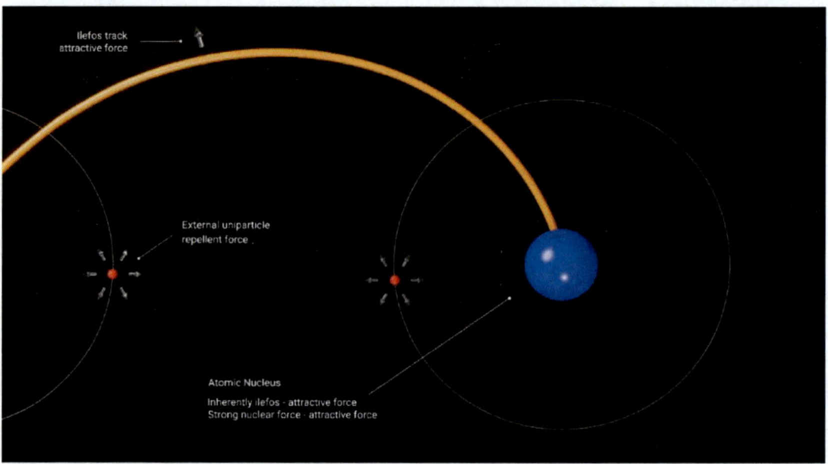


**Figure 8.** An atomic binding between two atoms and uni particles keeping the distance between the atoms (Stig Bø/Mohammed Ismaiel).

# Chapter 9

# Atomic Phase Displacement

An atom performs several actions and tasks. To perform these tasks the atom needs a specific energy for the task to be performed.

Each atomic quark is filled with a special energy, but often the atom does not have enough of the energy needed. The atom might have an abundance of another energy. The atom then converts this energy to the energy it needs. The conversion of energies in an atom is called Atomic Phase Displacement.

Each type of atom has special quark composition. This composition determines what energies and action the atom prefers.

If an atom has several plus quarks (ilefos, the attractive force, Higgs boson) in the neutron, it will prioritize ilefos and the attractive force if it has abundance of energy. This creates stronger ilefos tracks (atomic bindings/ gravity). We see this in iron.

If an atom has several energy quarks in the proton, it will prioritize dielectricity if it has abundance of energy. We see this in copper.

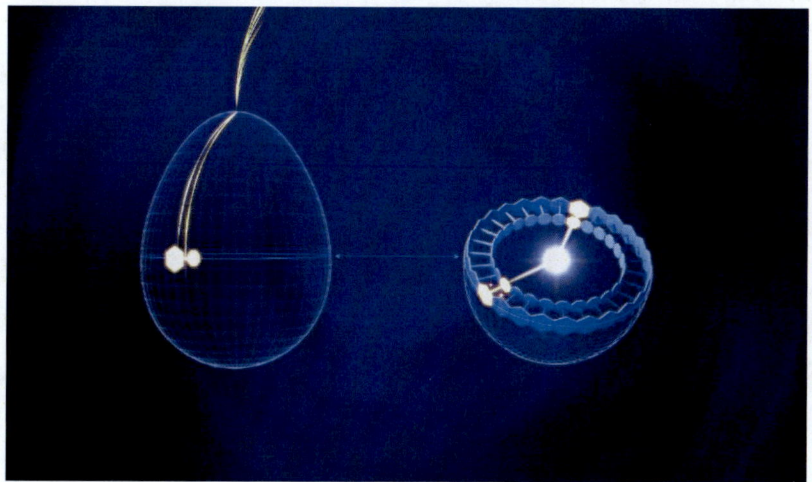

**Figure 9.** Quarks performing Atomic Phase Displacement in an atomic element (Ola Tandstad).

If we send strong dielectricity to iron, the iron has to get rid of the surplus energy. Iron converts this energy to its preferred energy, ilefos. It sends out strong ilefos tracks. Only atoms with similar quark composition can receive the full strength of attractive force. Other atoms with weaker ilefos tracks can't receive more ilefos (attractive force) than their tracks allow. Therefore only similar atom can feel this extra attractive force. This extra strong ilefos track is called magnetism. Only magnetic materials can feel this extra force. Magnetic material are atoms with similar quark composition. An electromagnet is an example of iron converting dielectricity to strong ilefos (magnetism).

If strong pulses of magnetism are received by atoms with quark composition which prefer dielectricity, the atoms convert this to dielectricity. Copper is such a material. Copper convert magnetism to strong dielectricity which it sends out in strong dielectricity tracks. This is a generator.

This is some examples of Atomic Phase Displacement. But Atomic Phase Displacements are being performed all the time in the atom. The strong energy, dimensional energies, the repulsive force and other energies are constantly being converted in an atom.

# Chapter 10

# Ilefos – The Attractive Force

Ilefos is energy units which attract each other. It has also a weak attractive force towards the strong energy.

When released from the neutron, these energy units form a track, due to the attractive force between the units. This track is an ilefos track, a gravity track.

An ilefos track will have attractive force towards external ilefos tracks from other atoms. This creates attraction between atoms.

Since the ilefos energy units also have a weak attraction towards the strong force, the units have to be sent out with a high speed to overcome this attraction. Due to the attraction between ilefos and the strong force, the ilefos track will not go in a straight angle out, but have a slight bend towards the nucleus. The track will straighten up with distance from the strong force (nucleus).

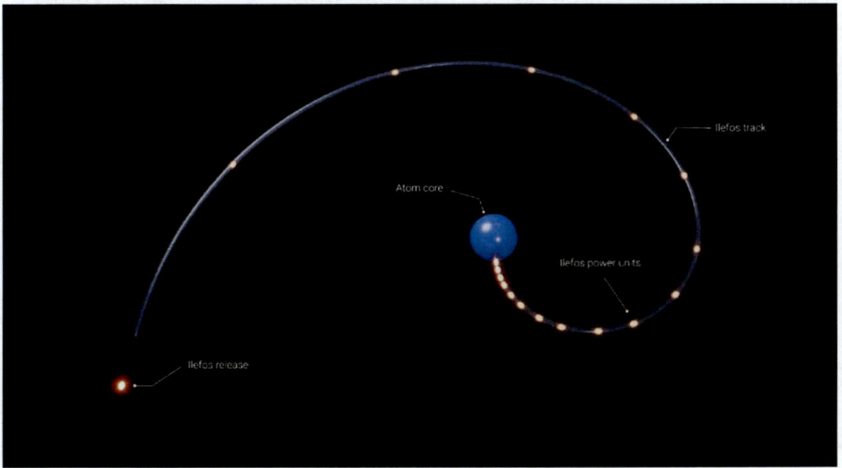


**Figure 10.** Ilefos energy units forming an ilefos (gravity) track from a neutron (Stig Bø/Mohammed Ismaiel).

After released from the neutron, the ilefos energy units form an ilefos track. The speed of the ilefos units in the track will gradually increase with distance from the strong force. When the speed of the units increases, the distance between the energy units in the track will gradually increase. With increased distance between the ilefos energy units in the ilefos track, the attractive force of the track will be reduced. The ilefos track is therefore strongest when emitted from the neutron, and the track will then have reduced attractive force with distance from the atom.

An ilefos track is therefore strongest close to the source (neutron/atom). When a strong ilefos connects with a strong external ilefos track, we will have a strong attraction, an atom binding.

With distance from the atom, the ilefos track become weaker. When two weaker ilefos tracks meet, we will have a weaker attraction, gravity.

The speed of the ilefos energy units in the track increases to the ilefos' energy speed. When the ilefos energy units are close to this speed, the distance between the energy units in the track become long. With long distance between the units, the attractive force between the units become weaker. At the end the attractive force between the units is not enough to keep the units in a track. The track will be dissolved into "free" ilefos energy units. This is called *ilefos release*. "Free" energy units, are energy units which are not connected to matter, particle or tracks, are dark energy. Around 68% of the universe is dark energy (Lambda-CDM model). When ilefos tracks and other energy tracks dissolve, they add dark energy to the universe. Free energy units are dark energy. Read more of this in the chapter "Dark energy."

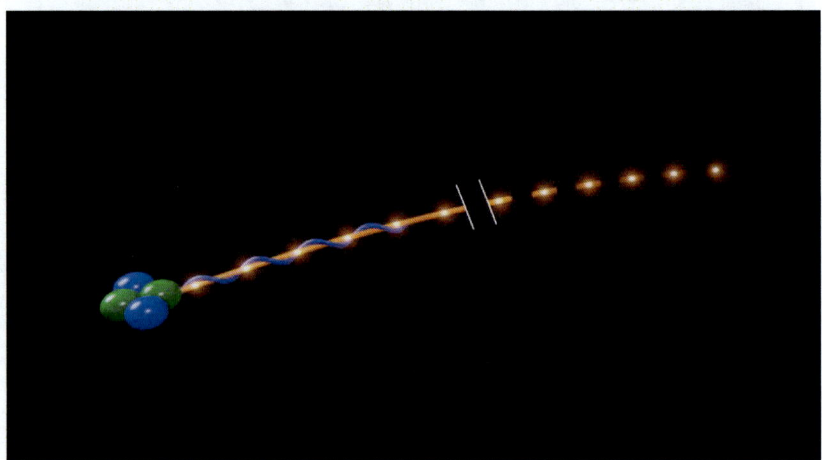

**Figure 11.** An ilefos track from an atom dissolve into free ilefos energy units (Stig Bø/Mohammed Ismaiel).

Ilefos energy units are released from the atoms. In normal hydrogen, which only has a proton, the tasks of the neutron are performed by the proton. The ilefos track from a hydrogen atom is therefore much weaker than from other atoms. This gives hydrogen very weak gravity. When two ilefos tracks meet, the attraction between the atom is determined by the weakest track.

The strength of the ilefos track is determined by the quark composition of the neutron. Plus quarks (Higgs bosons) are responsible for storing and converting ilefos, the attractive energy. If the neutron has only one plus quark, the neutron produces a weak ilefos track. In the periodic system, gases are atoms with only one plus quark, thus have weak ilefos tracks.

Neutrons with two plus quarks produce stronger ilefos track. These atoms might be characterized as solid.

Neutrons with three or more ilefos tracks produce strong ilefos tracks. These are metals in the periodic system.

The general energy level to the atom also affects the strength of the ilefos track. In an atom with very low energy level, the strength of ilefos tracks will be reduced. We see this in very cold matter, which have weaker atom bindings which are easily broken.

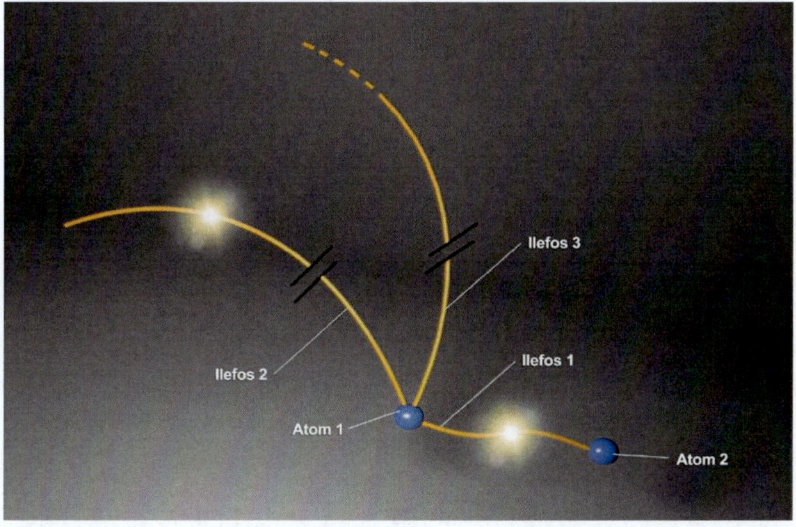


**Figure 12.** Three types of ilefos tracks: 1-Atomic binding, 2-gravity and 3-no external connection (ilefos release) (Stian Kongsvik).

To sum up:

Neutrons release ilefos energy units. These have attractive force towards each other and form a track. The strength of the track is strong close to its origin. The track loses power with distance from the source. The energy units accelerate, increase speed, which make longer distance between the units. This gives weaker attraction between the units and fewer units per distance track. When strong ilefos tracks meet close to source, they create a strong connection, an atomic binding. When weaker ilefos tracks meet further away from source, they create a weaker connection, gravity. With longer distance from the source, the ilefos track will become very weak. At the end the distance between the ilefos energy units will be too long, and they will not be able to hold each other together in a track. We will then have ilefos release. In ilefos release the ilefos track will dissolve and releases the ilefos energy units. The ilefos energy units become "free" energy units which are not connected to matter. Dark energy is "free" energy units. "Free" ilefos energy units are a significant part of dark energy.

## Graviton

In quantum gravity, graviton is a hypothetical elementary particle which mediates the force of gravity. In string theory some suggest graviton is a massless state of a fundamental string. None of these theories work with general relativity.

Present theories can't reconcile quantum mechanics with astrophysics. Ilefos power units and the ilefos model is a new approach to create an explanation which works in both quantum mechanics and astrophysics. An ilefos power unit is a new explanation of a gravity unit. A graviton is explained as a particle. An ilefos power unit is an ilefos energy pulse. In the ilefos model ilefos power units replace gravitons to explain gravity and atomic bindings.

# Chapter 11

# Uni Power – The Repulsive Force

Ilefos, the attractive force, pull atoms toward each other. To prevent collision between the nuclei of atoms, atoms also have a repulsive force. The repulsive force help atoms keep distance from each other, which prevent collisions of the atomic cores. This also help position atoms in a fixed position and help stabilize the ilefos tracks (prevent them from spinning).

I call the repulsive force for uni power. This force is stored and handled by TE quarks.

An atomic core can't send out both attractive force (ilefos tracks) and a repulsive force. These would neutralize each other.

To solve this, atoms have separate particles with repulsive force (uni power). These particles are a part of the atomic core, but are located outside the nuclei. Atoms have uni particles which orbit the atomic core. These particles have strong repulsive power towards external uni particles which orbit other atoms.

## Uni Particles

Uni particles form a protective "shield" around the nucleus. The particles orbit the nucleus at some distance. If the nucleus is 1 cm (0.4 inch) in diameter, the uniparticle will orbit 800 meters from the nucleus.

The uni particles also have a repulsive force towards ilefos tracks. This will force the ilefos tracks to be positioned between the uni particle orbits. The uni particles thus stabilize the ilefos tracks and prevent them from spinning. This helps the atoms to have stable attractive force and help keep the atoms in a fixed position.

Each neutron controls an uniparticle. The number of uniparticles are always equal to the number of neutrons. Again, with the exception of hydrogen which normally do not have a neutron. In hydrogen, the proton control of the uni particle.

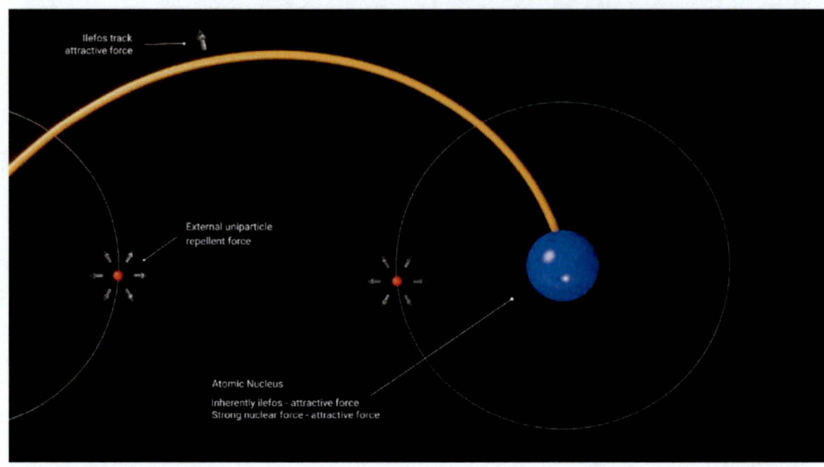


**Figure 13.** Uni particles performs the repulsive force to keep distance between two atoms (Stig Bø/Mohammed Ismaiel).

The neutron communicates and controls the uni particle. The neutron also converts uni energy from its TE quarks and send this energy to the uni particle.

The uni particles are separate particles with their own miniature management system, which in turn is controlled by the neutron. The uni particles consist of 10-15 quarks, which is only around 1/25 of the size of a neutron. A neutron can vary in size (number of quarks). Heavy/ larger neutrons have stronger ilefos tracks (attractive force), and therefore must have larger uni particles (larger repulsive force).

# Chapter 12

# The Strong Nuclear Force

The strong nuclear force holds the quarks and the nuclear elements (protons/ neutrons) together in one structure. This is the fundament of atoms.

The atoms constantly require a lot of energy to perform the strong nuclear force. YT quark in protons and neutrons stores, converts and handles YT energy, which performs the strong nuclear force. This might be gluon from the standard model.

YT quarks (gluons) store, release and convert this special energy. The strong nuclear force builds and maintains the structure of a quark. The same force also builds and maintains the structure of an atomic element. An atomic element, like a proton, is several quarks held together in a structure. The atomic quark can manage and run the operating system in the element together with support quarks.

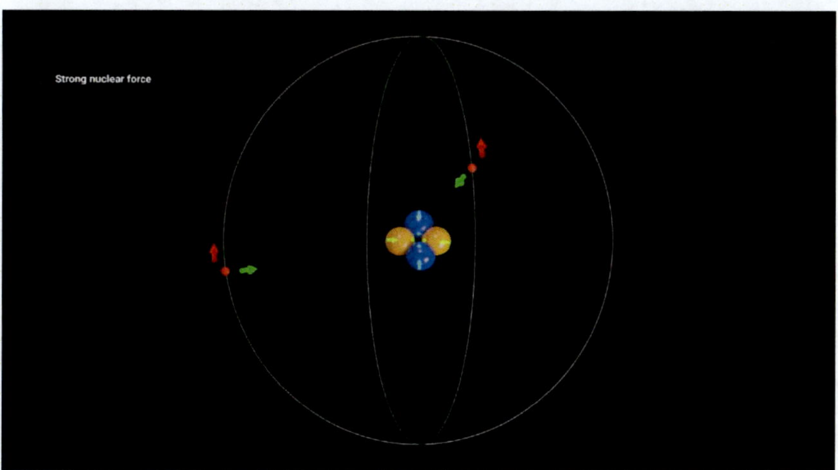


**Figure 14.** The strong nuclear force keeps the atomic elements together and hold the uni particles in orbits (Stig Bø/Mohammed Ismaiel).

The strong nuclear force performs a very strong force to keep the structures of an atom together. This is the strongest force of an atom. But it is sometimes not strong enough to keep the atomic elements together in the structure if two atomic cores (nuclei) collide. A collision between nuclei might knock one or more atomic elements loose from the nuclei. A very strong collision may destroy the whole structure, terminate the atom. We call this atom dissolution. It is therefore crucial for atoms to prevent collisions with their uniparticles and the repulsive force. Read more of this in chapter "Atom dissolution."

## Chapter 13

# Dielectricity

Through Atomic Phase Displacement the atom constantly converts energies to perform its tasks. The preferred energy for conversions is dielectricity. Dielectricity is the basic energy for energy transfers and energy conversion.

Other energies can also be converted in Atomic Phase Displacement, but they require more energy to convert. Dielectricity is pure energy.

All atoms and particles have a dielectricity quark, an energy quark. This quark may have different capacity. In the smallest particles like photons, the energy quark is minimum charged and is most basic.

In larger particles the energy quark grows in size and charge. Atoms is the first quark composition with fully functional atomic elements and a fully functional operating- and management system. An atom also has fully function energy quarks.

Hydrogen is the smallest element with a full atomic element. The proton in hydrogen has all the atoms' quarks and an atomic quark managing the atom's energies and other tasks.

Dielectricity is not called electricity because it is not transferred to and from atoms through particles (electrons). If your house is to share energy with other houses, you don't run around and share batteries. You use energy cables between the houses to transfer and share energy. Atoms also use the most efficient way to transfer and share energy. They also have energy tracks where they transport dielectricity. Thise are called dielectricity tracks.

Each proton produces a dielectricity track. A dielectricity track is dielectricity energy units released from the proton. Dielectricity energy units are pure energy units. They have a weak attraction to each other. When the units are released from the proton, the mutual attraction make the energy units form a track, a dielectricity track.

The dielectricity track are also attracted to ilefos and the ilefos track (gravity track). When released from the proton, the dielectricity track connects with the ilefos track and follow the ilefos track.

When an ilefos track connects with an external ilefos track and creates a binding, the dielectricity tracks also create a binding between the atoms.

Through this binding of the dielectricity tracks, atoms can transform dielectricity between the atoms.

As with an ilefos track, the dielectricity track become “weaker” with distance from source (atom). The speed of the dielectricity energy units will gradually increase in the track. This will increase the distance between the dielectricity energy units in the track, thus weaken the attraction between the energy units.

The ilefos track (gravity track) becomes also weaker with distance from source. When the ilefos tracks becomes very weak, the attractive force of the ilefos track will not be able to hold the dielectricity track. The dielectricity track will separate itself from the ilefos track and becomes a separate, free dielectricity track.

Short time before this separation, the dielectricity units will have an increased accelerating in speed. This will increase the distance between the dielectricity energy units, and the attraction between the units will become too weak for them to hold together in a track. The track will dissolve into free dielectricity energy units. Free dielectricity units are energy units which are not connected to matter or particles. These are a part of the dark energy composition.

The dissolution of the track is called dielectricity release.

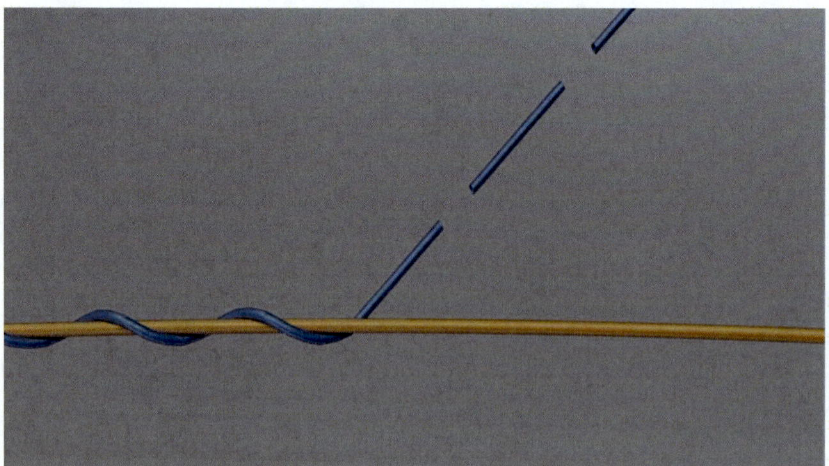

**Figure 15**. Dielectricity release (Stian Kongsvik).

## Dielectricity Capacity of Atoms

Different atoms have different dielectricity handling capacity. The number of energy quarks in the proton determines the energy (dielectricity) capacity.

An isolator has low energy handling capacity. These atoms have only one energy quark.

Copper is a metal with high electrical conductivity. This atom has three energy quarks and can have strong dielectricity tracks.

The number of energy tracks in relation to other quarks, like plus quarks, determines which energy the atom prefer when it has abundance of energy. Copper prefers to convert excess energies to dielectricity and transport them away in dielectricity tracks. Iron has four plus quarks (ilefos quarks), and prefer to convert excess energy to ilefos and transport them away in ilefos tracks.

Atoms have a preferred energy balance. This is when all atomic quarks are almost full of energy. Atoms with several energy quarks can store more dielectricity than atoms with fewer energy quarks.

Radioactive elements also are atoms with several energy quarks in their protons. They also have more protons than non-radioactive elements. What make them radioactive is that they also have more dimensional quarks than non-radioactive atoms. This makes them produce more dimensional energies.

See quark periodic table (Qperiodic table) in a later chapter for the quark composition of atoms in the periodic system.

## Chapter 14

# Dimensional Energies

Atoms communicate with each other through special energies. These energies are not affected by gravity and they therefore travel much faster than the speed of light.

The energies of the operating system and the internal management of an atom, are special energies which also are not affected by gravity (ilefos).

The energies which are not affected by the main forces to an atom, the strong nuclear force, ilefos (gravity), dielectricity and uni energy (the repulsive force), are dimensional energies. These energies are used in internal operations in atoms which are not related to these tasks. Also communication between atoms is performed by "neutral" dimensional energies.

Dimensional energies experience no drag from ilefos or dark energy, and travel very fast. They often form their own channels, a dimensional energy network, which make atoms able to communicate with each other.

There are three main types of dimensional energies:

1. Operating energies
2. Communication energies
3. Dimensional energies for information handling

Operating energies are dimensional energies which runs the atomic operating system and perform internal atomic tasks.

Communication energies are dimensional energies which atoms use for external communication. There are two groups of communication energies.

Short range communication energies are used for local communication in the vicinity with local atoms in the surroundings. Atoms have extended communications with their atomic neighbours. Atoms share information with each other to share energies and be aware of the surroundings.

Long range communication energies are dimensional energies which work at long range. These energies have separate communication network which extend all over the universe. This allow information handling energies to be in contact with all atoms and supervise them.

Dimensional energies for information handling are energies which survey and supervise all atoms and universal phenomena. These energies provide the atomic DNA, the recipe for how to build particles and atoms, which give the rules of energy behaviour. Our rules of physics are result of this dimensional energy.

When atoms experience to much dimensional energies, more than their dimensional quarks can handle, this can disturb the atomic operating system. This can make atoms behave in an unusual way. When cells change behaviour due to radioactive radiation (cancer), this is disturbance of the atomic operating system due to atoms being exposed to too much dimensional energies. These energies are hard to convert in Atomic Phase Displacement if the dimensional quarks do not have extra capacity.

## Chapter 15

# Three Types of Quarks

We have three types of quarks:

- Atomic Quark (management quark)
- Nuclear quarks
- Special quark

### Atomic Quark

Atomic Quark is the management quark. This quark run the atomic operating system. It is one atomic quark in every atom. This quark is responsible for every action performed by the atom. This is the brain of the atom. It communicates with nuclear quarks responsible for each atomic element.

### Nuclear Quarks

Nuclear quarks are "the workers" of the atom. A nuclear quark is responsible for handling a special energy (store, release, convert), or a special task in the local atomic element (proton/neutron).

Specific nuclear quarks can be responsible for a special energy, like plus quark (ilefos) and energy quark (dielectricity).

A specific quark can also be responsible for a special task in the atomic element, for example:

- Negative quark – responsible for internal communication in the element.
- Alfa quark – run the local operating system in the element – communicates with the atomic quark.
- Neutral quark – responsible for Atomic Phase Displacement (energy conversions) in the atomic element.

We also have several other nuclear quarks responsible for other energies and tasks.

## Special Quarks

Special quarks are the software quarks. These hold the instruction or program for the other quarks.

The atomic quark is connected to a special quark which hold the atomic DNA and the operating system. The atomic quark is the "server", the special quark connected to it holds the software (operating system).

Every nuclear quark is also connected to a special quark which hold the local software for the nuclear quark. Each nuclear quark has a special quark attached to it.

Together these quarks form the atom. Different types of atoms can have different numbers of nuclear quarks. The quark composition determines the properties of the atom. Atoms with three or more plus quarks (ilefos), have strong attractive force. These are characterized as metals in the periodic system. Atoms with only on plus quark, has weak attractive force. These are characterized as gases. The number of atomic elements (protons/neutrons), characterize also the atom. In the periodic system, each atom is described by the number of protons it has.

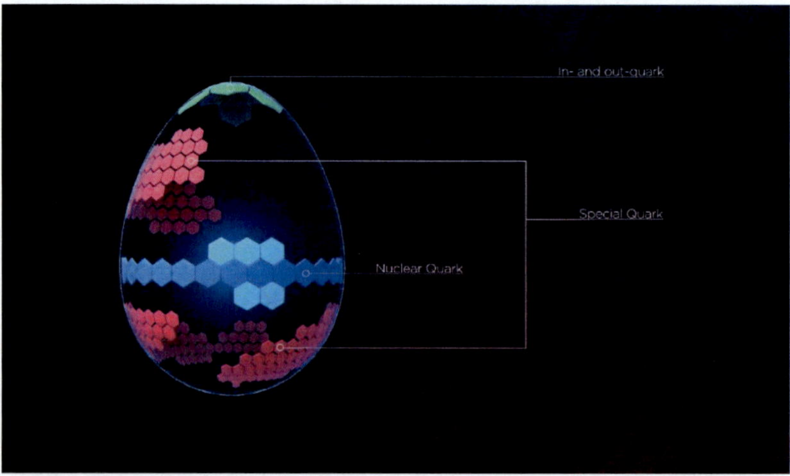


**Figure 16.** Different type of quarks in an atomic element (Ola Tandstad).

# Chapter 16

# An Atomic Operating System

We know atoms convert energies. We know atoms act on information from their surroundings to protect their structure. We know atoms may change properties if exposed to strong energies (isotopes). Atoms seem to communicate with each other, be aware of their surroundings, and share energies to have an even energy level. We must assume they also perform other tasks.

Like an advanced machine, atoms need commands to perform these tasks. They need a centralized management system, an operating system.

The atoms' operating system must be complex to handle all quarks, protons, neutrons and uniparticles. It must most likely have several specialized subsystems.

To run the atoms, each atom must have an atomic operating system. This system is managed by a quark, the atomic quark.

# Chapter 17

# Atoms as an Organism

Atoms continuous perform several complex tasks. Converting energies, communicating, act on information to preserve the atomic structure, hold together and control all quarks and atomic elements, are some of the main tasks.

The atomic quark is responsible for running the atomic operating system. We must assume the atomic quark is present in the preliminary stage in groups of particles before the atom is formed. I call these quark assemblies Resa.

If the atomic quark is present in preliminary stages, which can grow to atoms, and atoms can grow to larger atom (grow more atomic elements, protons/neutrons), and then finally grow to dark matter, the atomic quark must have all this information from the first quark assembling.

The atomic quark must have the blueprint for all atomic variations, have an atomic DNA.

I know DNA means deoxyribonucleic acid which is related to organisms. It carries genetic instructions for the development, function, growth, and reproduction to organisms. Atoms seem perform the same tasks. I therefore use the term "Atomic DNA" even though we here do not talk about an acid.

If the atomic quark can be seen upon as a server, it must also be connected to a special quark containing the instructions. The special quark connected to the atomic quark must provide the software/information. The special quark connected to the atomic quark can be called the atomic DNA and the software. The atomic quark is the management system acting with this special quark.

What characterizes an organism? An organism is a biological living system that functions as an individual life form. Do atoms also have a complex system which resembles the system of an organism?

Atoms seem to be aware of their surroundings and are able to act on this information to preserve their structure and share energies. The atomic quark which runs the atomic operating system, continuously communicates with all quarks and atomic elements, and keep them together in a fixed structure. To balance all these actions, the atoms continuously convert energies. If the conditions are right, the atom also may grow and change to an isotope or a new atom.

The actions of atoms seem to fulfil the definition of an organism. I therefore think we shall look on atom in an organism metaphor. Atoms might be looked upon as organisms.

# Chapter 18

# Energy Balance in Atoms

Atoms constantly use a lot of energy to perform their tasks. They use different types of energies to perform different types of actions. If the atoms do not have enough of a special energy, they convert other energies to the needed energy in Atomic Phase Displacement.

Atoms constantly use energies. But to be in energy balance, they must receive the same amount of energy. How can the atoms receive energy?

An atom can receive energy through:

1. From other atoms' energy tracks (sharing energy)
2. From external energetic quarks
3. From free energy units

Energy received from other atoms through energy tracks, is dielectricity received through dielectricity tracks. Each proton has a dielectricity track. Through a dielectricity track an atom can receive or release dielectricity energy units. The dielectricity tracks is the preferred way to receive energy or dispatch surplus energy. Atoms can also receive energy from ilefos tracks.

An atom can absorb free energetic quarks. If an atom's energy tracks do not have capacity to transport away surplus energy, they can produce and dispatch energetic quarks. Also, universal phenomena produce and release energetic quarks, which might get together in small quarks groups. An atom can absorb these free quarks and quark groups.

An atom constantly receives free energy units through its "antennas." Free energy units are energy units which are not connected to matter or universal phenomena like stars or black holes. The fabric of space is filled with free energy units. Dark energy, 68% of the visible universe, are free energy units. These travels through space until they are absorbed by atoms or universal phenomena.

When particles, quarks, or atoms dissolve, they release their energies as free energy units. When ilefos- or dielectricity tracks become too weak to stick together, they dissolve and release their energy as free energy units. This is the main sources to dark energy (free energy units).

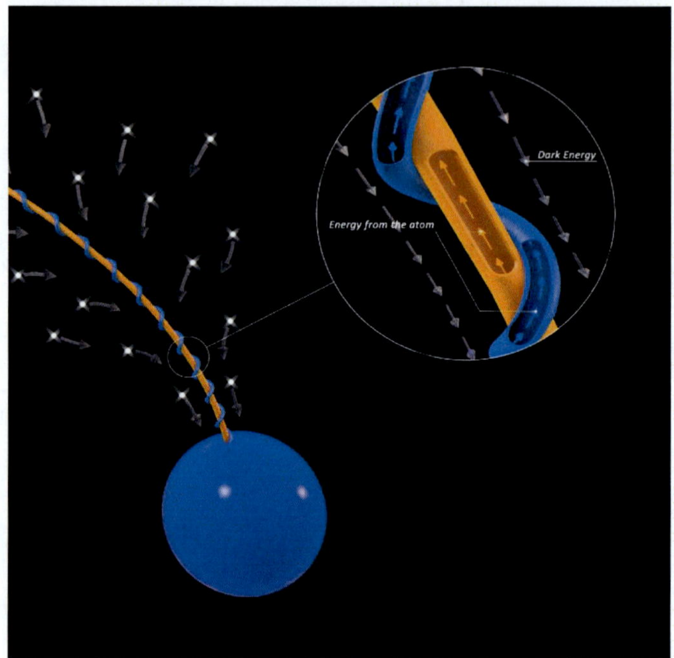


**Figure 17.** An atom harvests dark energy (free energy units). The energy units go to the atom. At the same time the atom release energy in their ilefos- and dielectricity tracks (Ola Tandstad).

Through ilefos- and dielectricity tracks atoms send out ilefos- and dielectricity energy units in energy tracks. These tracks also attract similar energy units due to the attraction between energy units. Free energy units which the tracks come in contact with or are in the vicinity of the tracks, are drawn towards these tracks. They are then lead through the outside of these tracks towards the atomic core and absorbed by the nucleus.

The ilefos- and dielectricity tracks are also the antennas which harvest free energy unis.

The harvesting of free energy units is the main way an atom receive energy. The length and strength of the tracks determines how much free energy units an atom can receive. The concentration of free energy units in the area also determines the feed of free energy units.

Energy units are then recycled in atoms.

Universal phenomena also recycle energy units when they dissolve matter, particles, and quarks and release their energy units. These energy units are also released in energy tracks and may form quarks and particles, like photons, when they travel away from the source.

The atom constantly uses a lot of energy performing its tasks. The atom also constantly receive energy from external energy tracks, from external particles/ quarks, and from harvesting free energy units (dark energy) through its harvesting tracks (ilefos- and dielectricity tracks).

The atom has a preferred energy level where all quarks are almost full, and their tasks are of optimum effect.

If the atom receives too little energy, the atom must lower the level of some of its tasks. To save energy and be in energy balance, the atom first reduces the power of their ilefos- and dielectricity tracks. In cold conditions, the atom therefore send out less heat (dielectricity) and attractive force (atomic bindings). We see this when extremely cold materials are more likely be more brittle due to weaker atomic bindings. They also have a reduce weight. (This is also according to $E = mc^2$. In the special relativity a mass weighs more when heated).

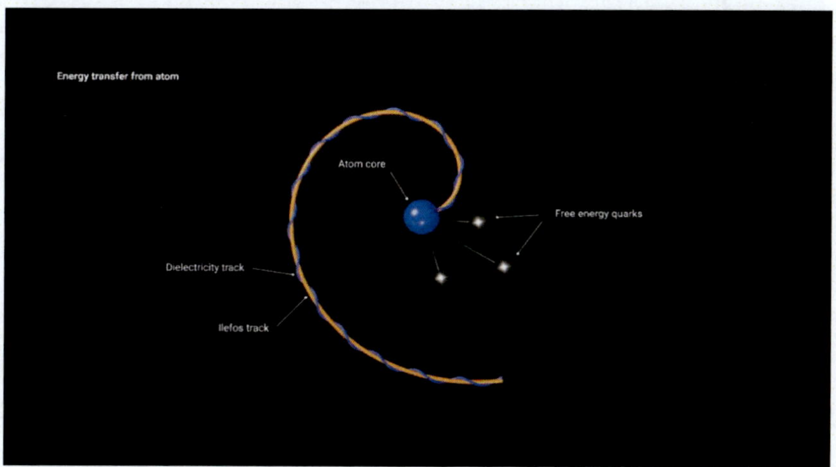


**Figure 18.** Energy transfer from an atom through ilefos- and dielectricity tracks and through release of quarks/photons. (Stig Bø/ Mohammed Ismaiel).

If the atom receives more energy than it needs and can store in its quarks, the atom will first try to get rid of the surplus energy through its energy tracks. If the atom can't get rid of this energy when these tracks are of maximum effect, the atoms start to produce energetic quarks, energy quarks, which it releases. These quarks might be gathered in quark groups like photons. The free quarks which do not form photons, will soon after release dissolve into free dielectricity units, which by living organisms are perceived as heat.

If the atom suddenly receives extremely large amount of energy, the atom might produce so many quarks that they form new atomic elements (protons/neutrons). The atom then grows in size. We see this in special universal phenomena like in a supernova or in strong energy rays from black holes (Hawking radiation) or other strong energy rays.

## Chapter 19

# The Nature of Energy and Gravity Tracks

The attractive force becomes weaker with distance from source. Most energy and other forces have the same properties – they become weaker with distance from source. Why?

Isaac Newton law of universal gravitation states that every particle attracts every other particle in the universe. This force is proportional to the product of their masses and inversely proportional to the square of the distance between their centres:

$$F = G \frac{m_1 m_2}{r^2}$$

**Figure 19.** Isaac Newton's law of universal gravitation.

F is the gravitational force acting between two objects, $m_1$ and $m_2$ are the masses of the objects, r is the distance between the centres of their masses, G is the gravitational constant.

This is an alternative to general relativity. In Newton's formula the attraction between masses is a force which decreases with distance. Then both atomic bindings and gravity can be the same attractive force.

This is a mathematical calculation, or description of observations. What cause the attractive force and why do it decrease in strength with distance?

The attractive force I call ilefos. Ilefos energy units attract each other.

When the ilefos energy units are released from a proton, the attractive force between the units makes them form a track. I call this an ilefos track. The attractive force of the ilefos units make this a gravity track which attract external ilefos tracks from other origins.

The ilefos energy units are also attracted by the strong nuclear force in the atom. Therefore, the ilefos energy units must be shot out with a speed high enough to fight the attraction of the strong nuclear force. If the units are released with a too slow speed, the tracks will return to the source after a short while.

**Figure 20.** An ilefos energy unit (Stig Bø/Mohammed Ismaiel).

Due to the attraction to the strong nuclear force, the ilefos track has a Fibonacci like form. After released, the ilefos energy units will increase in speed. When their speed increase, the distance between each unit will increase, thus it will be fewer ilefos energy units at a set distance of an ilefos track.

With fewer ilefos energy units per set distance, the attraction to the strong nuclear force will decrease. This will make the ilefos track slowly to straighten out, thus have a fibonacci form. With some distance from the atom, the ilefos track will travel in an almost straight line.

A connection between two ilefos tracks close to the atom, where they are strong, is an atomic binding. When two ilefos tracks connects further out, they are weaker and make a weaker connection. The weaker connection is gravity.

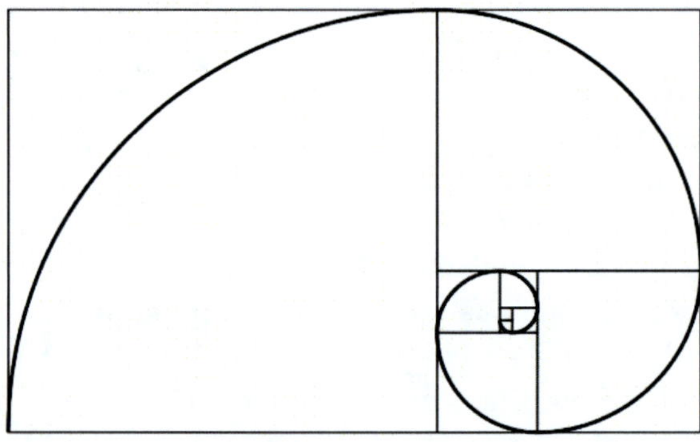

**Figure 21.** Fibonacci spiral (Dicklyon).

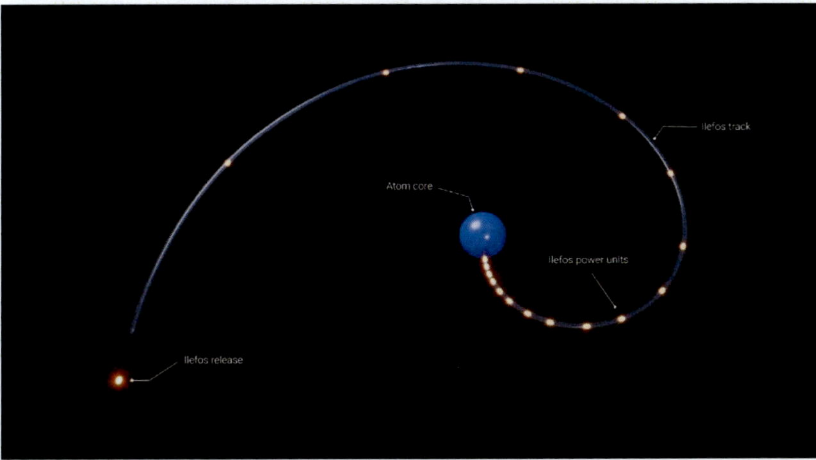


**Figure 22.** Ilefos energy units form an ilefos track when released from the neutron. The units accelerate and this create longer distance between the units (Stig Bø/Mohammed Ismaiel).

The attraction of the ilefos track will with distance from the atom slowly decrease as the distance between the ilefos energy units increases.

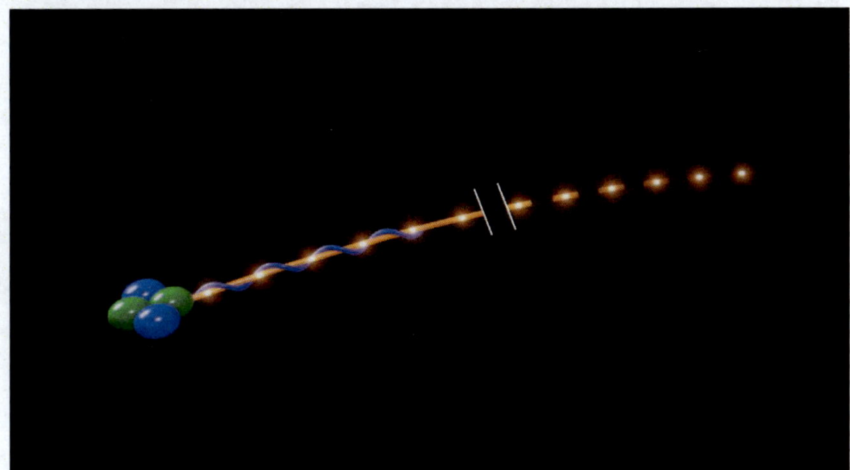

**Figure 23.** Ilefos release (Stig Bø/Mohammed Ismaiel).

The speed of the energy units will slowly increase until they reach their energy speed. The energy speed is the natural speed the energy units travel in. When ilefos energy units are approaching this speed, the distance between the units is very long. Then the attraction between the energy units will not be strong enough to keep the units in a track. The ilefos track will dissolve into free energy units which are not connected to each other. This is ilefos release. Free energy units are dark energy.

This behaviour we see in all energy units released from atoms or universal phenomena which have attraction to each other. Most similar energy units have attraction towards each other.

Dielectricity energy units also attract each other and form a track when they are released from the proton. Dielectricity energy units are also attracted to ilefos, so the dielectricity track turns around the ilefos track after release from the atom. The dielectricity track also become weaker with distance from source, the dielectricity energy units increase in speed.

At some distance the dielectricity track will become too weak to hold on to the also weaken ilefos track. The dielectricity track will the separate from the ilefos track. Soon after the dielectricity track will not be able to hold together, and the dielectricity track will dissolve into free dielectricity energy units. I call this dielectricity release.

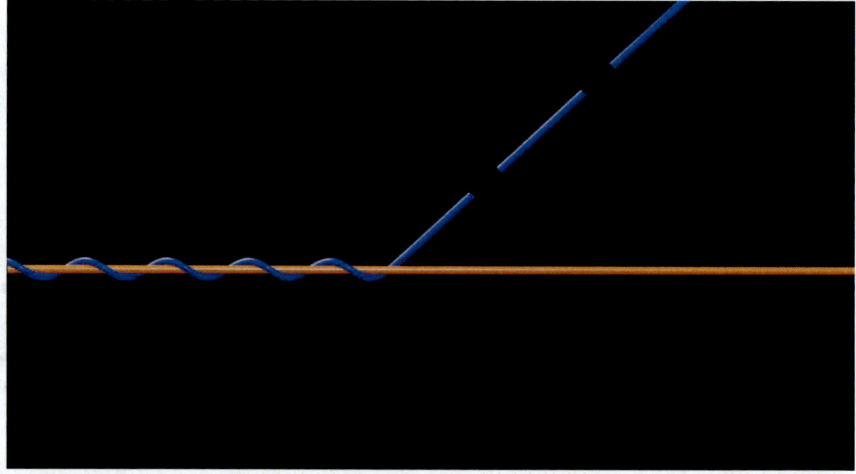

**Figure 24.** Dielectricity release (Stian Kongsvik).

## Chapter 20

# The Weakening of Atomic Bindings

Atomic bindings are when ilefos tracks from different atoms connect close to the atom where the ilefos tracks are stronger. A permanent connection between strong ilefos tracks close to the atoms are called an atomic binding.

The strength of the binding is determined by the weakest tracks. When two ilefos tracks of equal strength connects, we have optimum strength. We see this when iron atoms connect and form atomic bindings. They then form strong atomic bindings.

Gasses have weaker ilefos tracks. When two gasses' ilefos tracks meet, we will have a weak atomic binding. We also have a weak binding if a gas connects to a metal's ilefos track.

Atomic bindings of gasses are weak. $CO_2$ for example, have weak atomic bindings. When the oxygen atom receives extra energy, it will produce stronger uni particles (repulsive force). This pushes the other atom further away, thus increase the length of the ilefos tracks in the atomic binding. Longer ilefos tracks become weaker. The weaker binding with the other atoms will break. This is what happens in photosynthesis. The energic oxygen atom will connect to an external ilefos track of another oxygen atom, which also have a strong ilefos track, forming $O_2$. The carbon will together with water ($H_2O$) form hydrocarbon in the plant.

In combustion or burning of hydrocarbons, oxygen release energy to the hydrocarbon. The hydrocarbon' uni particles then orbit in wider orbits, which break the atomic bindings to the hydrogen. The result from the combustion is $CO_2$ and water ($H_2O$).

Metals have stronger ilefos tracks and stronger atomic bindings. When a metal receives extra energy, the atoms will also produce stronger uni particles and repulsive force. This will also push the atoms away from each other.

The longer distance between the atoms in the metal, will give longer and weaker atomic bindings between the iron atoms.

When iron is heated (receive energy), the atomic bindings become weaker and iron glow and can be shaped. If further energy (heat) is given to the atoms, the atomic bindings will be further weakened. The metal then will melt. If extremely high energy is introduced, the atomic binding will weaken to a degree that the atomic bindings are broken. The iron atoms will then be "free" atoms with no atomic bindings to other iron atoms.

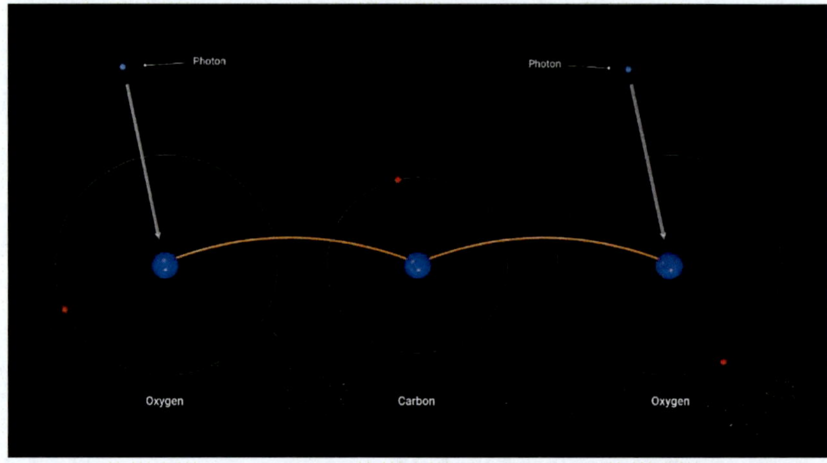


**Figure 25.** $CO_2$ receives more energy from photons (Stig Bø/Mohammed Ismaiel).

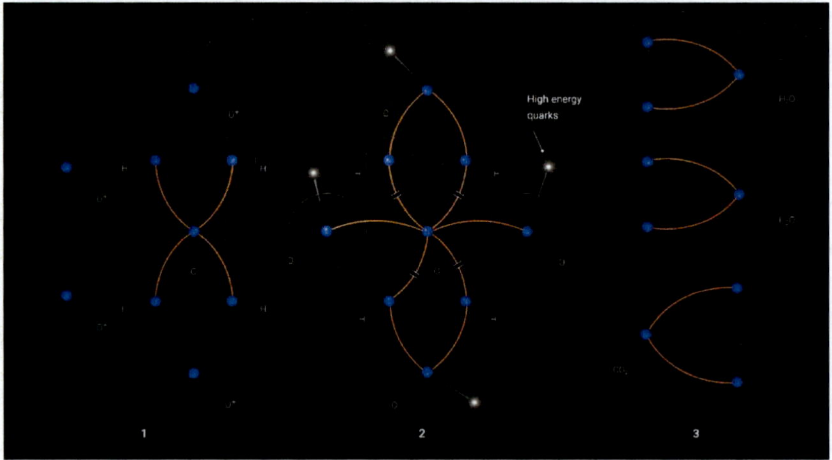


**Figure 26.** The combustion process (Stig Bø/Mohammed Ismaiel).

This physics we see in all atoms. The ilefos tracks have limited maximum strength. The strength of the ilefos tracks of an atom is determined by the number of plus quarks (ilefos quarks) in the neutron.

If the atom receive extra energy, more than the quarks can handle, the atom must get rid of the extra energy. The atom's ilefos tracks are already at maximum strength, so it has to remove the excess energy through other channels. First it increases the dielectricity tracks to their maximum strength. Increased dielectricity might be perceived as heat. Then it increases the strength of the uni particles, the repulsive force increases. This pushes the atomic core further away from each other. If this is not enough, they produce energy rich quarks which they release. These free quarks might form photons (light) and be absorbed by other atoms. This makes the atom produce heat and light.

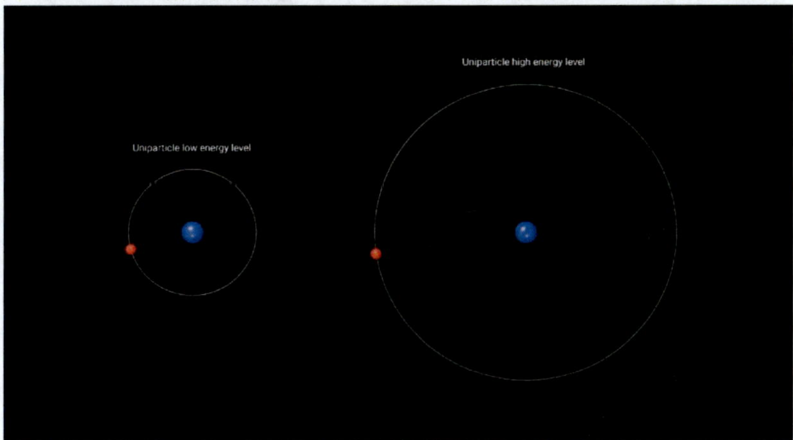


**Figure 27.** Uni particles with different energy levels (Stig Bø/Mohammed Ismaiel).

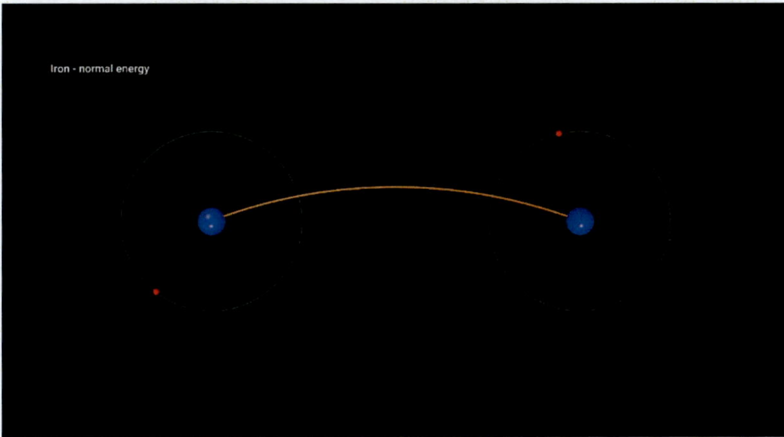


**Figure 28.** An iron atomic binding (Stig Bø/Mohammed Ismaiel).

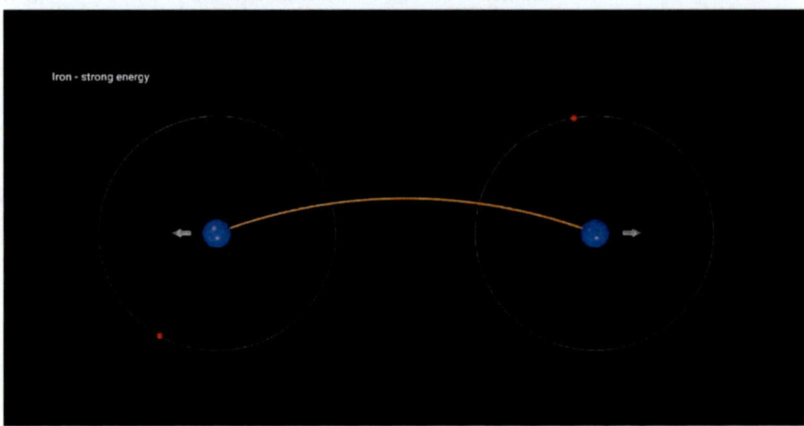


**Figure 29.** The uni particles receive extra energy, find wider orbits and push the atoms further away from each other (Stig Bø/Mohammed Ismaiel).

**Figure 30.** Melted iron (Colourbox).

If atoms fast receive extremely large amount of energy, the atom will produce many quarks, and this quark production will be so intensive that the atom produce new atomic elements (protons and neutrons). This will be described in a later chapter.

## Super Conductor

When the atom has very low energy, they do not send out dielectricity tracks. They must conserve the energy to the primary tasks. Two of the most important tasks are the strong nuclear force and the repulsive force.

The attractive force is not that important when the atom shortage of energy. The attractive force is reduced. The ilefos tracks are then weaker, and the ilefos energy units are sent out with a lower force and speed. The ilefos tracks are sent out from the neutron, but they are not strong enough to overcome the strong nuclear force of the atom. The ilefos track will travel out, and then be drawn back to the neutron. The atom will then have few or no outgoing ilefos tracks. The atom might still have ilefos tracks which form weak atomic bindings to neighbour atoms, but no going out to perform gravity.

We can see this in supercooled materials which might hover due to lack of gravity. Dielectricity tracks have a weak attraction towards ilefos and usually twist around the ilefos track.

**Figure 31.** An atom with very low energy in ilefos tracks - A super conductor (Stian Kongsvik).

When strong dielectricity tracks travel through supercooled material, the dielectricity tracks do not connect to the very weak ilefos tracks. They go straight forward without connecting with the atoms. The dielectricity track then do not experience the resistance of going through atoms. When dielectricity tracks go through a material with almost no resistance, we call this material super conductor.

**Figure 32.** A levitating super conductor (Julian Litzel).

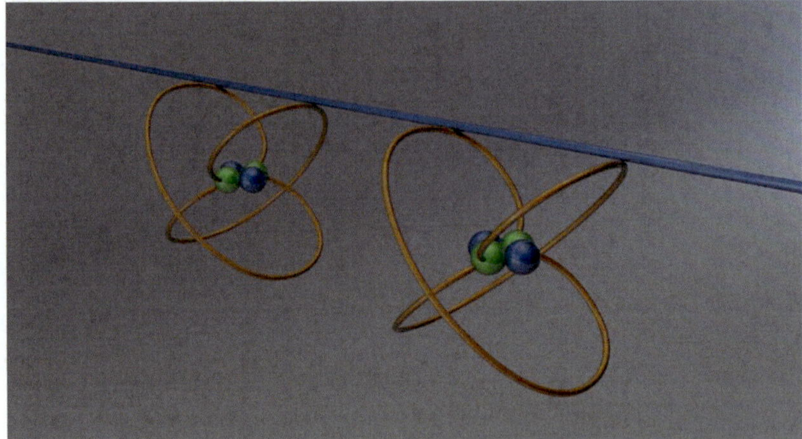

**Figure 33.** Super conductor atoms have too weak gravity tracks to attract a dielectricity track (Stian Kongsvik).

# Chapter 21

# Particles and How Matter and Atoms Are Created and Evolve

Quarks can't survive by itself without a control quark. Without control the quarks will not receive the strong force which hold them together. The quarks will then within a short time dissolve and release their energy.

Groups of quarks may connect and form a primitive operating system. When two quarks connect, they automatic form an atomic quark which manages the group. The smallest group of quarks are of three quarks: An energy quark, an ilefos quark and an atomic quark.

## Photons

The smallest groups of quarks are photons. A photon consists of 1-4 energy quarks, 1-2 ilefos quarks and one atomic quark. Each of these also have a special quark connected to itself, with the software/ instructions to the main quark.

A star sends out ilefos- and dielectricity tracks. It also sends out a lot of quarks, primarily energy quarks. These quarks might connect and form groups of quarks like photons.

## Resa

Sometimes they also form larger groups of quarks. We see this in the sunin coronal mass ejections (CME). This is groups of quarks which are larger than photons, but smaller than atoms. I call these resa.

Resa are groups of quarks which might be called particles. They have evolved to a more complex structure with more types of quarks. Resa might have a weak uni particle, but they do not have ilefos- or dielectricity tracks.

## Resa 0

Resa 0 is the simplest form of resa. It has between 10 and 50 quarks, including special quarks. It has no ilefos track or uni particles. It is a group of quarks which are held loosely together through the build-in gravity of ilefos quarks. Resa 0 is very weakly affected by gravity, has weak energy and no uni particles. It still lacks many of the nuclear quarks of an atom.

## Resa

Normal resa has between 51 and 160 quarks and has a more complex composition than resa 0. Resa starts to look like a proton, but it still lacks many of the nuclear quark of the proton. It has evolved a TE quark and has made a weak uniparticle which help protect its structure. It still does not have ilefos- and dielectricity tracks, but it is more complex than resa 0. It has more energy and are held stronger together. It is more affected by gravity than resa 0.

## Resa+

Resa + has between 161 and 255 quarks. It is denser and has almost all the quarks of an atom. Resa + can grow into nearly a proton. It has a fully functional uni particle, but still not enough nuclear quarks to form ilefos- and dielectricity tracks. It remains resa until it has all types of nuclear quarks of an atom (all 23 types).

## Atom

When the group of quarks grow to 256 quarks, the group has all 23 types of nuclear quarks. It then evolves into a fully functional quark group with a full operating system. It has one or more complete protons. The atom has uni particle (s) to protect the structure (repulsive force), and one or more ilefos- and dielectricity tracks. The atom converts energies and communicates with other atoms, it has a strong nucleus and structure.

## How Particles Evolve to Matter (Atoms)

Particles like photons or resa can evolve (grow) by:

1. Absorbing other particles
2. Grow by receiving external energy

If a group of quarks, like resa, have a weak collision with another particle, like a photon, the largest group of quarks will attract and absorb the weakest group. In this process the atomic quark of the weakest group dissolves, and the quarks will be managed by the largest groups' atomic quark. The atomic quark might form a basic Atomic Phase Displacement and change the quarks to the most needed quarks. This is growth by absorption.

If a group of quarks receive strong energy from an energetic dielectricity track or another strong energy track, the quark group will not have capacity to store this energy in its existing quarks. The atomic quark then builds new quarks to store this energy. This is growth by internal quark production.

When resa by one of these methods grow to a quark group with more than 256 quarks, it converts into a fully atom with a full atomic operating system.

Atoms can also grow. Atoms grow by fusion and internal growth.

If two atoms travel in the same direction, and at the same time are exposed to strong energy, the uniparticles will become energetic and obtain wide orbits. If the two atoms' uni particles orbits cross each other, the nuclei are not protected from collision.

A strong collision might crush the nuclei or part of the nuclei.

A soft collision will not destroy the nuclei, but make the nuclei merge. The largest nucleus will absorb the smallest nucleus. The neutrons and protons of the smallest atom will be controlled by the strongest nucleus' atomic quark and become a part of its structure. The atomic quark of the weakest atom will dissolve.

An Atom can also grow when exposed to a strong energy ray. Supernovas, black holes (Hawking radiation) and other universal phenomena can produce strong energetic rays. When an atom is exposed to such strong energy, it is not able to store this energy in its quarks. It has not capacity to get rid of the excess energy through its normal tracks/ways. It has to produce quarks to store the energy. When it produces too many quarks, the atom does not have capacity to expel them. The atom then forms new atomic elements, make new protons and neutrons. Atoms can in this way evolve and grow to larger atoms with more atomic elements.

## Dark Matter

When an atom receives more than 291 quarks in each atomic element, it also starts making new types of quarks. Normal matter has 23 types of quarks and has between 256 and 290 quarks in each atomic element.

When an atom grows to more than 290 quarks and more than 23 types of quarks in the atomic elements, it becomes dark matter.

Dark matter is an evolution of matter. Dark matter has the same properties as matter, but with enhanced actions.

Dark matter has several more energy quarks and has therefore much larger energy capacity. It has capacity to absorb all the energy of photons, thus not reflect light. It has much longer and stronger gravity tracks, which can harvest much more dark energy to the nucleus. This is because it has several more plus quarks (ilefos quarks) than normal atoms. Dark matter also has many TE quarks which produce very strong uni energy (repulsive energy). Dark matter has therefore very large and strong uni particles in very wide orbits around the nucleus. This causes a much larger distance between the nuclei of dark matter than ordinary matter. The absorbing of all photons and the large distance between the dark matter nuclei, make dark matter hard to observe visually. We can though identify areas with dark matter through the strong gravity.

# Chapter 22

# Atomic DNA

If atoms and particles can grow, evolve, they must have a blueprint, a recipe how to grow. All matter in the universe seems to have the same recipe. The periodic system lists up the known atoms.

I call this recipe the atomic DNA. DNA is not the correct term. DNA mean deoxyribonucleic acid and give genetic instructions for development and function of living organisms and viruses. Atoms seem to have the same recipe. I therefore adapt this term and call it Atomic DNA.

When the smallest groups of quarks form, they also seem to create a management quark. I call this quark the atomic quark. This quark seems to follow the quark group as it grows, providing instructions for growth and structure of the group.

Photons, resa, atoms and dark matter seem to keep the same atomic quark. If two groups of quarks meet and fusion, they both have an atomic quark. The new larger group of quarks only need one atomic quark. The atomic quark of the smallest group then dissolves.

When an atomic quark is formed, this quark will handle and execute actions based on the recipe of the atomic DNA. The atomic DNA contains the instructions and the software to the atomic quark. The atomic DNA is a special quark working with the atomic quark. The atomic quark is the server of the atom (or quark group), the special quark connected to the atomic quark contains the software for the atomic quark.

The atomic DNA contains all instructions for the structure of all sizes of quark groups. When a quark group grow, either through fusion or internal growth, the atomic DNA have the instructions for the new structure.

The atomic DNA also have instructions for other actions and behaviour of the quark group. When the group is large enough, it will form an atomic element, a proton. It then contains all quarks it needs to have a fully operational atomic operating system. It then performs the actions of an atom:

- It produces forces like the strong nuclear force, the attractive nuclear force (ilefos), the repulsive force (uni power), and communication energies.

- It constantly converts energies in Atomic Phase Displacement to balance and perform these tasks.
- It communicates with other atoms and is aware of its surroundings.
- It receives energies from other atoms and dark energy.
- It shares energies with other atoms.
- If it receives abundance of energy, it releases excess energy through energy tracks and may produce and release energetic quarks.
- If it receives very strong energy, it might produce enough quarks to form a new atomic element (proton/neutron).
- If it has a soft collision with another atom, the nuclei might fusion to a larger atom.
- If it has a hard collision and the nucleus is damaged, loose quarks or an atomic element, it will use energy trying to produce new quark and replace the missing quarks/element.

The atomic quark and its special quark (atomic DNA), work together to form a fully functional operating system of the atom or group of quarks.

A special quark is a quark containing the software or instructions. Nuclear quarks are quarks operating special energies or performing special tasks in the atom. Each nuclear quark has a special quark containing the software or the quark's instructions. The atomic quark is the atom's operating system which operate the atom through nuclear quarks.

# Chapter 23

# Dark Matter Atoms

When ordinary matter receives large amounts of energy, they can produce quarks to get rid of the excess energy. If an atom produces enough quarks, it will use the quarks to produce new atomic elements (proton/neutrons). The atom then grows in size to a new atom with more protons and neutrons.

If atoms travel in the same direction and receive large amounts of energy, their wider uni particles' orbits may cross, and we can have a soft nuclei collision. This can lead to a fusion of the atom. This also make an atom grow in mass.

Ordinary atoms have between 256 and 290 quarks in their atomic elements. They consist of a mix of 23 different nuclear quarks.

If an atom reaches the end of our periodic table (+5 protons), it reaches the limit for mass of ordinary atoms. If the atom then grows through one of the methods mentioned, we see a dramatic change.

The atom will then build larger atomic elements, but also produce a new type of nuclear quark (UR energy, a dimensional energy). It will also have several more energy- and plus quarks (ilefos), which give the atom increased energy handling capacity. It can absorb and store much more energy. The ilefos (gravity) tracks also increase in strength and length. Longer ilefos tracks can harvest more dark energy. Increased strength and length of the ilefos tracks give much stronger gravity.

This would result in very strong atomic bindings and a dense material, if it was not for the increase power to the uni particles (repulsive force). When dark matter atoms have more energies, their uniparticles also increase in size and strength. The more powerful uni particles will travel in wider orbits further away from the dark matter nucleus. The powerful uni particles will push other dark matter atoms away, and we will have a large distance between the dark matter atoms. The increased orbits and strength of the uni particles makes the nuclei keep a distance which is 500-1000 times the distance to an ordinary atom.

Dark matter might therefore be hard to spot due to the wide structure.

The increased energy handling capacity make a dark matter atom absorb all energy from photons, thus not reflect any light. This together with the longer distance between the dark matter atoms, make them hard to observe visually.

We can see the influence of dark matter on ordinary matter and stars, which are influenced by the strong and long gravity tracks of dark matter atoms.

The atomic quark is the same in dark matter atoms as in ordinary atoms. The atomic quark performs the same tasks, only much stronger. Dark matter has larger atomic elements with different quark composition, and a new type of nuclear quark, which give it stronger energy and dimensional properties.

The dark matter atoms may also grow in size up to the dark matter limit given by the atomic DNA.

Dark matter atoms can be absorbed by stars and black holes. They then dissolve and release their energy.

Large amount of dark matter can form dark matter fields. These might become very dense despite their strong uni particles. If the field is dense enough, the heat from uni particle touching (friction) might provide enough energy to ignite a star. This star will grow very fast in size due to the large amount of dark matter it can feed on.

We believe the universe consist of:

| Dark energy | 68,2% |
|---|---|
| Dark matter | 26,8% |
| Ordinary matter | 5% |

Dark matter then constitutes of around 85% of all mass. We can't observe dark matter, we only know it is there due to the interaction with ordinary matter through gravity.

85% of all matter is dark matter do not mean that 85% of the atoms are dark matter atoms. The dark matter atoms send out much stronger and longer gravity tracks compared to ordinary atoms. The average strength of ilefos tracks (gravity) in dark matter atoms are 200 times stronger than the ilefos track of an average ordinary atom. This means we have more ordinary atoms than dark matter atoms, but the stronger gravity to dark matter, make them produce 85% of the gravity in the universe.

- Dark matter has many more plus quarks than ordinary matter. It has more plus quarks than energy quarks, and prioritizes therefore ilefos (gravity) tracks over dielectricity tracks. It therefore does not emit heat.

- Dark matter has several strong TE quarks, and therefor produce large amounts of uni energy (repulsive force). Dark matter therefore has very strong uni particles which orbit the dark matter atom, which pushes other dark matter atoms away. They keep other dark matter atoms at a large distance from the nucleus. The distance between the nuclei are more than 500-1000 times the distance we find in ordinary matter.
- The many plus quarks produce very long and powerful ilefos (gravity) tracks. These produce great attractive force over long distances. They also harvest very much dark energy to the dark matter atom. Due to the many energy quarks all this energy can be stored and converted in the nucleus. The large energy capacity also makes dark matter absorb all energy of photons, thus not reflect light.

# Chapter 24

# The Altering Energy Levels of Atoms

An atom is usually in energy balance. This means the energy the atom uses to perform its tasks, is equal to what the atom constantly receives. The atom constantly receives energy from dark energy, loose particles like photons, and through energy tracks from other atoms. The atom tries to balance the incoming energy with its use of energy. The atom tries to keep their quarks at almost full energy level to keep the atom safe from external threats. The quarks are therefore almost always 95-100% full of energy.

But special atomic incidents may force the atom to change this energy balance to preserve the atomic structure.

## Friction

Atoms can release energy to protect their structure. Atoms have a repulsive force which make the atoms keep distance. This prevents the nuclei from collide with each other. A collision of nuclei might destroy parts or the whole atomic core. To prevent this, atoms have uni particles which orbit the nucleus. These particles have repulsive force towards other atoms' uniparticles, and help atoms be in a fixed position away from each other.

Atoms communicate with each other and are aware of their surroundings. If an atom travels or is forced toward another atom, the standard energy level of the uni particles might not be enough to keep the distance. To prevent a collision between the nuclei, the atom increases the production of the repulsive force (uni power). This is an extraordinary incident, and this forces the atom to use more energy than it constantly receives. The atom has to use stored energy from its quarks. This makes the atom use more energy than it receives, which bring it out of energy balance.

What happen next depends on the force of the collision:

1. Friction
2. Nuclei touching

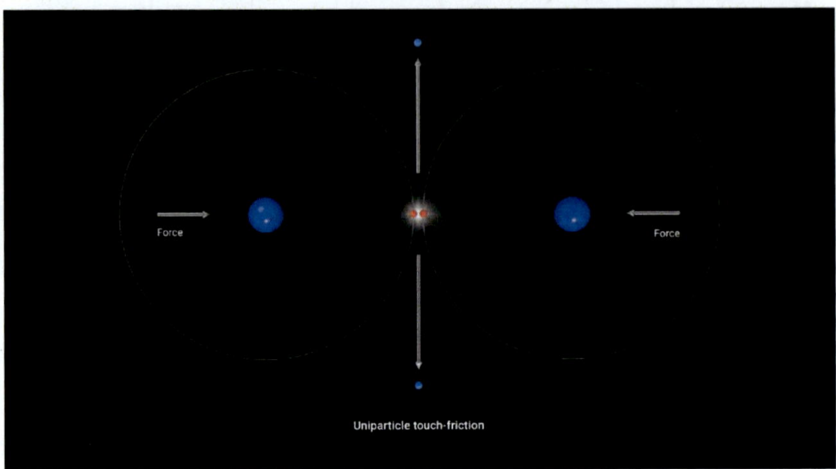


**Figure 34.** Uni particle touch - friction (Stig Bø/Mohammed Ismaiel).

Normally the increased repulsive power of the uni particles is enough to protect the nucleus. When two uni particles from different atoms come close, they release strong uni power. This leads to strong uni power in the vicinity. This causes other atoms to receive increased energy (uni power energy units), which they convert to dielectricity. This is perceived as heat.

If the repulsive force is not strong enough, the uni particles might touch each other. This result in a uni particle collision which destroy the uni particles which collides. The quarks in the uni particles are released as free quarks. After a very short while these free quarks will dissolve and release their energy. This results in a release of energy, which increases the concentration of dark energy (free energy units) in the area. Increased dark energy means atoms will experience increased energy income. They might not be able to store all this energy, and transport the excess energy away through dielectricity tracks and production of quarks which they release. This can be experienced as heat and light.

The atom will immediately make new uni particles and replace them. But this also make them use stored energy, bringing them out of energy balance.

The release of energy in friction depends on which type of friction we have. If the uni particles only release strong repulsive force but retain their structure, we have moderate release of energy. If the uni particles collide and are destroyed, we have a much stronger release of energy.

## Fission

If uni particles (the repulsive force) are not able to keep the nuclei away from each other, we will have nuclei touching or nuclei collision. The force of this collision determines what happen next.

A soft collision will make the external nucleus knock away an atomic element. One or more neutrons can be knocked away from the atomic core.

The released neutron will preserve its structure for a short while. But without an atomic quark, it will dissolve. When it dissolves, the neutron releases all its quarks. Theise quarks will also quickly dissolve and will release their energy. This leads to a strong increase of free energy units (dark energy) in the area.

The neutron knocked away from the atom, maintains its structure for a short while. If it survives long enough to hit another nucleus, it might knock away one or more neutrons from this atom. This might result in a chain reaction as we see in a nuclear bomb.

The atom which loose one or more neutrons, will immediately make new quarks and rebuild the neutron. This means that the atom must draw energy from its remaining quarks, and the energy level of the atom is reduced.

In time the atom will through energy tracks from other atoms and dark energy, return to the normal energy level of the atom.

## Atom Dissolution

An atomic collision where the nuclei collide, might destroy the whole structure of the atom. If the collision is weak, one or more atomic elements like neutrons, can be knocked away from the nucleus. But the atom is able to rebuild them and preserve the structure of the atom.

If the collision between nuclei is strong, this collision might knock loose too many atomic elements, like neutrons, and the atom will not have enough energy rebuild the elements. This leads to that the atom will have too little of the strong energy to keep the remaining of the atomic elements together. All atomic elements will be unleashed, and the atomic quark will disappear. All atomic elements (protons, neutron and uni particles), will become free elements with no atomic quark to control them. After a short while these elements will dissolve and release their quarks. These quarks will either form small quark groups or dissolve and release their energy. This results in an immense release of different types of energies. This is atom dissolution.

**Figure 35.** Atom dissolution (Stig Bø/Mohammed Ismaiel).

An atom dissolution also gives several free neutrons which might travel a short while before they dissolve. If some of these neutrons hit a nucleus, they might knock loose other neutrons, and we have a chain reaction. A H bomb (hydrogen bomb) an example of this.

- Matter grows when it is receives much energy. Atoms then can grow in mass.
- When this matter dissolves, it releases the same energy.

A Star constantly receives groups of quarks, matter, and dark energy. The star dissolves the quark groups and matter to release their energy.

This energy is released as free energy units. These energy units may form quarks when they are dense and travel in the same direction. These quarks might form small quarks groups like photons and resa like we see from our sun. The sun release photons (light) and may also release resa (particles) as we see in coronal mass ejection (CME).

The free energy units which attract each other, form dense tracks when leaving the star. When ilefos energy units form a track, this is a very dense and strong gravity track. Dielectricity energy units also form strong tracks which connects with the ilefos tracks. The dielectricity tracks are perceived as heat.

Black holes also suck inn matter and particles. These (matter and particles) are also completely dissolved in atom dissolution and all energy are released. The concentration of ilefos (gravity) energy units are in black holes so strong that they do not release dielectricity from the black hole. Except from the "top and bottom" of the whole where the gravity (ilefos concentration) is weaker. Here can dielectricity and other energy units escape. These might form quarks in the strong energy beam (Hawking radiation). This energy and these quark group might in turn start new stars.

Energy is recycled through matter, stars, and black holes.

## Fusion

Fusion is when two atoms merge and create an atom with more mass. This can only happen in specific conditions.

To have a fusion we must have a soft collision between two nuclei. But atoms have uni particles with repulsive force which help prevent a nuclei collision. This help atoms to preserve their structure. How can we have a soft collision between two nuclei?

If atoms receive large amounts of energy, they have to get rid of the excess energy. If they can't get rid of his energy through ilefos- and dielectricity tracks, they will produce energetic quarks which they release. At the same time, they also produce large amounts of uni energy, the repulsive force, in the uni particles. The uni particles, which go in orbits around the nucleus, have repulsive force towards other uni particles. In this way they help the atoms keep distance and this protects the nuclei.

When the uni particles become very energetic and filled with uni power, they go faster, in wider orbits and have increased repulsive force.

When the uni particles' orbits become very wide, the uni particle orbits might intersect with the orbits of uni particles from other atoms. If this happens, the uni particles can no longer protect the nuclei from touching each other.

If this happens when the atoms travel in the same direction, we might have a soft collision between the nuclei, a nuclei touch.

This touch is not strong enough to knock atomic elements loose. Instead, the two atomic cores will merge. The atomic quark of the smallest atom will dissolve, and the remaining atomic quark will take control of all atomic elements, thus forming a new atom containing all atomic elements of the two former atoms.

This process does normally not produce excess energy.

But if one or both atoms is an isotope with one or more extra atomic elements (neutrons), the new atoms will have an uneven number of protons and neutrons. The surplus element(s) are then released. This/ these elements will be free element(s), which within a short time will dissolve due to it/ they do not have an atomic quark to manage it. The free element(s) will dissolve, and the quarks will release their energy.

A fusion process therefore does not release energy, unless one (or both) of the atoms has extra neutron(s).

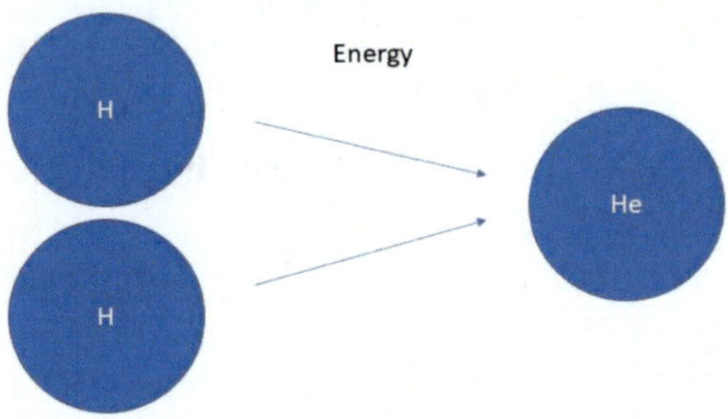


**Figure 36.** Fusion between two hydrogen atoms to a helium atom (Bent Rolf Pettersen).

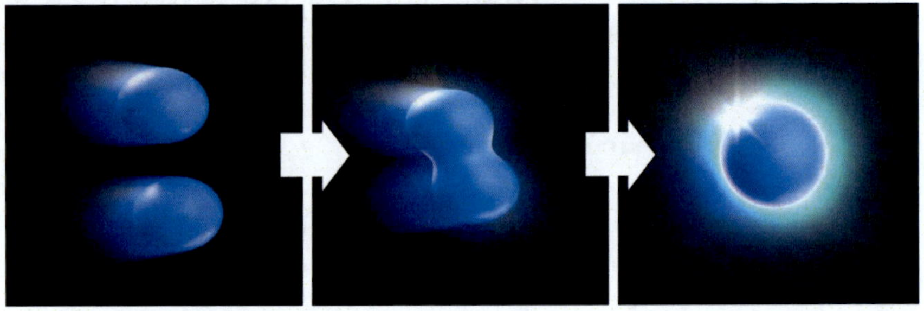

**Figure 37.** Fusion, a soft nuclei collision (Ola Tandstad).

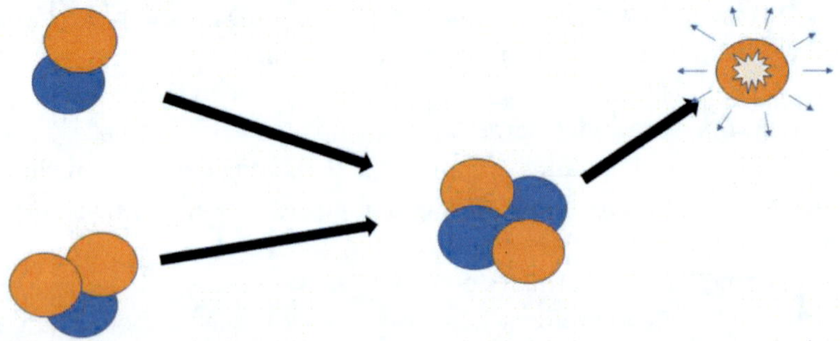

**Figure 38.** Fusion between atoms with an extra neutron. The extra neutron is released, dissolves and releases its energy (Bent Rolf Pettersen).

# Chapter 25

# The Periodic Table

The periodic table of the elements, arranges atoms into groups. The atoms are arranged in atomic numbers based on how many protons the atom have. The periodic table also try to sort the elements in groups with similar properties.

The first periodic table was made by chemist Dmitri Mendeleev in 1869 where atoms were sorted by the chemical properties of their atomic mass. Later the table evolved with the discovery of atomic numbers and other properties.

| Group→<br>↓Period | 1 | 2 | | 3 | 4 | 5 | 6 | 7 | 8 | 9 | 10 | 11 | 12 | 13 | 14 | 15 | 16 | 17 | 18 |
|---|---|---|---|---|---|---|---|---|---|---|---|---|---|---|---|---|---|---|---|
| 1 | 1 H | | | | | | | | | | | | | | | | | | 2 He |
| 2 | 3 Li | 4 Be | | | | | | | | | | | | 5 B | 6 C | 7 N | 8 O | 9 F | 10 Ne |
| 3 | 11 Na | 12 Mg | | | | | | | | | | | | 13 Al | 14 Si | 15 P | 16 S | 17 Cl | 18 Ar |
| 4 | 19 K | 20 Ca | | 21 Sc | 22 Ti | 23 V | 24 Cr | 25 Mn | 26 Fe | 27 Co | 28 Ni | 29 Cu | 30 Zn | 31 Ga | 32 Ge | 33 As | 34 Se | 35 Br | 36 Kr |
| 5 | 37 Rb | 38 Sr | | 39 Y | 40 Zr | 41 Nb | 42 Mo | 43 Tc | 44 Ru | 45 Rh | 46 Pd | 47 Ag | 48 Cd | 49 In | 50 Sn | 51 Sb | 52 Te | 53 I | 54 Xe |
| 6 | 55 Cs | 56 Ba | * | 71 Lu | 72 Hf | 73 Ta | 74 W | 75 Re | 76 Os | 77 Ir | 78 Pt | 79 Au | 80 Hg | 81 Tl | 82 Pb | 83 Bi | 84 Po | 85 At | 86 Rn |
| 7 | 87 Fr | 88 Ra | ** | 103 Lr | 104 Rf | 105 Db | 106 Sg | 107 Bh | 108 Hs | 109 Mt | 110 Ds | 111 Rg | 112 Cn | 113 Nh | 114 Fl | 115 Mc | 116 Lv | 117 Ts | 118 Og |
| | | | * | 57 La | 58 Ce | 59 Pr | 60 Nd | 61 Pm | 62 Sm | 63 Eu | 64 Gd | 65 Tb | 66 Dy | 67 Ho | 68 Er | 69 Tm | 70 Yb | | |
| | | | ** | 89 Ac | 90 Th | 91 Pa | 92 U | 93 Np | 94 Pu | 95 Am | 96 Cm | 97 Bk | 98 Cf | 99 Es | 100 Fm | 101 Md | 102 No | | |

**Figure 39.** The periodic table.

But this periodic table still do not explain why the elements have their properties. I therefore suggest a new periodic table where we also explain the basic quark composition of the atoms, the Quark Periodic System.

# Chapter 26

# Qperiodic System

Different type of atoms prefers different types of energies. The quark composition of the atom decides which type of energies it prefers, and how much of the energies it can store and convert.

The number of plus quarks determines how much ilefos an atom can store and convert. This determines how much of the attractive force an atom can release. The attractive force determines the strength of the atom bindings to the atom and its gravity.

- An atom with one plus quark, have low power in their ilefos track, and therefore low strength in their atom bindings. These atoms are gases. These atoms have only capacity to produce weak ilefos tracks and do not prefer ilefos in atomic phase displacement.
- An atom with two plus quarks, have moderate ilefos capacity. It can store and convert more ilefos energy and have moderate strong ilefos tracks. These elements have moderate strong atom bindings and are called solid elements.
- An atom with three or more plus quarks, have strong ilefos capacity. It can store and convert larger amounts of ilefos and can have strong ilefos tracks. It therefore has strong atomic bindings and is called metal.

Dielectricity is the preferred energy for sharing and the easiest energy to convert. An atom sends out dielectricity through dielectricity tracks. How much dielectricity an atom can store and convert depends on how many energy quarks the atom has.

- An atom with one energy quark, has low dielectricity capacity. It can store and convert a low amount of dielectricity. The atom therefore has weak dielectricity tracks. Due to its weak dielectricity tracks, the atom is called an isolator.
- An atom with two energy quarks has moderate dielectricity capacity and can have moderate strength in the dielectricity tracks.

- An atom with three or more energy quarks has high dielectricity capacity. This atom store larger amounts of dielectricity and can have very strong dielectricity tracks. Copper is such an atom. It conducts dielectricity very well and are therefore called a conductive atom.
- Some atoms also have extra dimensional quarks. The atom uses dimensional energies for communication internal and external. Dimensional energy also affects the atomic operating system. An atom with extra dimensional quark(s), have better capacity for storing and converting dimensional energy, and therefore release more dimensional energy than other atoms. An atom with extra dimensional quark(s) is called a radioactive element. These tend to be the heaviest elements, and they have many strong atomic elements with many quarks. They are the stage of atoms which can grow to dark matter.
- To take these factors into account, I have made a new periodic system which list up plus quarks, energy quarks and extra dimensional quarks in the elements. We then can have a better understanding of the properties of atoms.
- The *ilefos potential* is how much ilefos energy which can be stored in the neutrons. We find this when we multiply the number of plus quarks with the number of neutrons. (Protons also have plus quarks and can store ilefos. The total ilefos energy in the whole atom therefore have the double of this energy).
- The *dielectricity potential* is how much dielectricity energy which can be stored in the protons. We find this when we multiply the number of energy quarks with the number of protons. (Neutrons also have energy quarks and can store dielectricity. The total dielectricity energy in the whole atom therefore have the double of this energy).
- *Light dark matter* is when matter start to form extra dimensional quarks. These atoms are very large with better energy handling capacity than "normal" matter. They have many atomic elements (protons/ neutrons) and start to form more dimensional quarks when they grow. This is a sign for a coming big step in the next stage of atomic evolution. The growth of extra dimension quark is to prepare for a new type of dimensional quarks, which we only find in dark matter.
- To grow from light dark matter (radioactive elements) to dark matter require immense energy. This energy can't be found at earth. When

light dark matter is exposed to immense energy, the extra dimensional quarks are transformed to a new type of quarks which only can be found in dark matter. At the same time the atom grows several extra of plus quarks (ilefos), energy quarks and other quarks, transforming it to dark matter. Due to these quarks, dark matter has extremely high energy capacity. It absorbs alle energy from photons, and thus do not reflect light. The gravity of dark matter is also much greater than the gravity of matter.

## Qperiodic table

| Number atom elements * | Name | | Plus quarks | Energy quarks | other Dim. quarks | Category | Category 2 | Ilefos potential | Energy potential |
|---|---|---|---|---|---|---|---|---|---|
| 1 | Hydrogen | H | 0 | 2 | 0 | Gas | Matter | 0,1 | 1 |
| 2 | Helium | He | 1 | 2 | 0 | Gas | Matter | 2 | 4 |
| 3 | Lithium | Li | 2 | 2 | 0 | Light matter | Matter | 6 | 6 |
| 4 | Beryllium | Be | 2 | 1 | 0 | Light matter | Matter | 8 | 4 |
| 5 | Boron | B | 2 | 1 | 0 | Light matter | Matter | 10 | 5 |
| 6 | Carbon | C | 2 | 2 | 0 | Light matter | Matter | 12 | 12 |
| 7 | Nitrogen | N | 1 | 1 | 0 | Gas | Matter | 7 | 7 |
| 8 | Oxygen | O | 2 | 3 | 0 | Gas | Matter | 16 | 24 |
| 9 | Fluorine | F | 1 | 1 | 0 | Gas | Matter | 9 | 9 |
| 10 | Neon | Ne | 1 | 1 | 0 | Gas | Matter | 10 | 10 |
| 11 | Sodium | Na | 2 | 1 | 0 | Light matter | Matter | 22 | 11 |
| 12 | Magnesium | Mg | 2 | 2 | 0 | Light matter | Matter | 24 | 24 |
| 13 | Aluminum | Al | 3 | 2 | 0 | Light matter | Matter | 26 | 26 |
| 14 | Silicon | Si | 2 | 1 | 0 | Light matter | Matter | 28 | 28 |
| 15 | Phosphorus | P | 2 | 3 | 0 | Light matter | Matter | 30 | 45 |
| 16 | Sulfur | S | 2 | 3 | 0 | Light matter | Matter | 32 | 48 |
| 17 | Chlorine | Cl | 1 | 3 | 0 | Gas | Matter | 17 | 51 |
| 18 | Argon | Ar | 1 | 1 | 0 | Gas | Matter | 18 | 18 |
| 19 | Potassium | K | 2 | 3 | 0 | Light matter | Matter | 38 | 57 |
| 20 | Calcium | Ca | 2 | 2 | 0 | Light matter | Matter | 40 | 40 |
| 21 | Scandium | Sc | 2 | 1 | 0 | Light matter | Matter | 42 | 21 |

**Figure 40.** Qperiodic table -1 (Bent Rolf Pettersen).

**Qperiodic table**

| Number atom elements * | Name | | Plus quarks | Energy quarks | other Dim. quarks | Category | Category 2 | Ilefos potential | Energy potential |
|---|---|---|---|---|---|---|---|---|---|
| 22 | Titanium | Ti | 3 | 2 | 0 | Metal | Matter | 66 | 44 |
| 23 | Vanadium | V | 3 | 1 | 0 | Metal | Matter | 69 | 23 |
| 24 | Chromium | Cr | 3 | 2 | 0 | Metal | Matter | 72 | 48 |
| 25 | Manganese | Mn | 3 | 2 | 0 | Metal | Matter | 75 | 50 |
| 26 | Iron | Fe | 4 | 2 | 0 | Metal | Matter | 104 | 52 |
| 27 | Cobalt | Co | 3 | 3 | 0 | Metal | Matter | 81 | 81 |
| 28 | Nickel | Ni | 3 | 1 | 0 | Metal | Matter | 84 | 28 |
| 29 | Copper | Cu | 3 | 3 | 0 | Metal | Matter | 87 | 87 |
| 30 | Zinc | Zn | 3 | 1 | 0 | Metal | Matter | 90 | 30 |
| 31 | Gallium | Ga | 3 | 1 | 0 | Metal | Matter | 93 | 31 |
| 32 | Germanium | Ge | 3 | 1 | 0 | Metal | Matter | 96 | 32 |
| 33 | Arsenic | As | 3 | 1 | 0 | Metal | Matter | 99 | 33 |
| 34 | Selenium | Se | 3 | 1 | 0 | Metal | Matter | 102 | 34 |
| 35 | Bromine | Br | 3 | 1 | 0 | Metal | Matter | 105 | 35 |
| 36 | Krypton | Kr | 1 | 1 | 0 | Gas | Matter | 36 | 36 |
| 37 | Rubidium | Rb | 3 | 1 | 0 | Metal | Matter | 111 | 37 |
| 38 | Strontium | Sr | 3 | 1 | 0 | Metal | Matter | 114 | 38 |
| 39 | Yttrium | Y | 3 | 1 | 0 | Metal | Matter | 117 | 39 |
| 40 | Zirconium | Zr | 3 | 1 | 0 | Metal | Matter | 120 | 40 |
| 41 | Niobium | Nb | 3 | 1 | 0 | Metal | Matter | 123 | 41 |
| 42 | Molybdenum | Mo | 3 | 1 | 0 | Metal | Matter | 126 | 42 |
| 43 | Technetium | Tc | 3 | 1 | 0 | Metal | Matter | 129 | 43 |
| 44 | Ruthenium | Ru | 3 | 1 | 0 | Metal | Matter | 132 | 44 |
| 45 | Rhodium | Rh | 3 | 1 | 0 | Metal | Matter | 135 | 45 |
| 46 | Palladium | Pd | 3 | 1 | 0 | Metal | Matter | 138 | 46 |
| 47 | Silver | Ag | 3 | 1 | 0 | Metal | Matter | 141 | 47 |
| 48 | Cadmium | Cd | 3 | 1 | 0 | Metal | Matter | 144 | 48 |
| 49 | Indium | In | 3 | 1 | 0 | Metal | Matter | 147 | 49 |

* Number of atom element: Neutron, Proton

**Figure 41.** Qperiodic table – 2 (Bent Rolf Pettersen).

| Number atom | Name | | Plus quarks | Energy quarks | Other dim. | Category | Category 2 | Ilefos potential | Energy potential |
|---|---|---|---|---|---|---|---|---|---|
| 50 | Tin | Sn | 3 | 1 | 0 | Metal | Matter | 150 | 50 |
| 51 | Antimony | Sb | 3 | 1 | 0 | Metal | Matter | 153 | 51 |
| 52 | Tellurium | Te | 3 | 1 | 0 | Metal | Matter | 156 | 52 |
| 53 | Iodine | I | 3 | 1 | 0 | Metal | Matter | 159 | 53 |
| 54 | Xenon | Xe | 1 | 1 | 0 | Gas | Matter | 54 | 54 |
| 55 | Cesium | Cs | 3 | 1 | 0 | Metal | Matter | 165 | 55 |
| 56 | Barium | Ba | 2 | 1 | 0 | Light matter | Matter | 112 | 56 |
| 57 | Lanthanum | La | 2 | 1 | 0 | Light matter | Matter | 114 | 57 |
| 58 | Cerium | Ce | 2 | 1 | 0 | Light matter | Matter | 116 | 58 |
| 59 | Praseodymium | Pr | 2 | 1 | 0 | Light matter | Matter | 118 | 59 |
| 60 | Neodymium | Nd | 3 | 2 | 0 | Metal | Matter | 180 | 120 |
| 61 | Promethium | Pm | 2 | 1 | 0 | Light matter | Matter | 122 | 61 |
| 62 | Samarium | Sm | 2 | 1 | 0 | Light matter | Matter | 124 | 62 |
| 63 | Europium | Eu | 2 | 1 | 0 | Light matter | Matter | 126 | 63 |
| 64 | Gadolinium | Gd | 2 | 1 | 0 | Light matter | Matter | 128 | 64 |
| 65 | Terbium | Tb | 2 | 1 | 0 | Light matter | Matter | 130 | 65 |
| 66 | Dysprosium | Dy | 2 | 1 | 0 | Light matter | Matter | 132 | 66 |
| 67 | Holmium | Ho | 2 | 1 | 0 | Light matter | Matter | 134 | 67 |
| 68 | Erbium | Er | 2 | 1 | 0 | Light matter | Matter | 136 | 68 |
| 69 | Thulium | Tm | 2 | 1 | 0 | Light matter | Matter | 138 | 69 |
| 70 | Ytterbium | Yb | 2 | 1 | 0 | Light matter | Matter | 140 | 70 |
| 71 | Lutetium | Lu | 2 | 1 | 0 | Light matter | Matter | 142 | 71 |
| 72 | Hafnium | Hf | 3 | 1 | 0 | Metal | Matter | 216 | 72 |

**Figure 42.** Qperiodic table -3 (Bent Rolf Pettersen).

| Number atom * | Name | | Plus quarks | Energy quarks | Other dim. | Category | Category 2 | Ilefos potential | Energy potential |
|---|---|---|---|---|---|---|---|---|---|
| 73 | Tantalum | Ta | 3 | 1 | 0 | Metal | Matter | 219 | 73 |
| 74 | Tungsten | W | 3 | 1 | 0 | Metal | Matter | 222 | 74 |
| 75 | Rhenium | Re | 3 | 1 | 0 | Metal | Matter | 225 | 75 |
| 76 | Osmium | Os | 3 | 1 | 0 | Metal | Matter | 228 | 76 |
| 77 | Iridium | Ir | 3 | 1 | 0 | Metal | Matter | 231 | 77 |
| 78 | Platinum | Pt | 3 | 1 | 0 | Metal | Matter | 234 | 78 |
| 79 | Gold | Au | 3 | 1 | 0 | Metal | Matter | 237 | 79 |
| 80 | Mercury | Hg | 3 | 1 | 0 | Metal | Matter | 240 | 80 |
| 81 | Thallium | Tl | 3 | 1 | 0 | Metal | Matter | 243 | 81 |
| 82 | Lead | Pb | 3 | 1 | 0 | Metal | Matter | 246 | 82 |
| 83 | Bismuth | Bi | 3 | 1 | 0 | Metal | Matter | 249 | 83 |
| 84 | Polonium | Po | 3 | 1 | 0 | Metal | Matter | 252 | 84 |
| 85 | Astatine | At | 3 | 1 | 0 | Metal | Matter | 255 | 85 |
| 86 | Radon | Rn | 1 | 2 | 0 | Gas | Matter | 86 | 172 |
| 87 | Francium | Fr | 2 | 1 | 0 | Light matter | Matter | 174 | 87 |
| 88 | Radium | Ra | 4 | 3 | 1 | Radiactive matter | Light dark matter | 352 | 264 |
| 89 | Actinium | Ac | 3 | 2 | 1 | Radiactive matter | Light dark matter | 267 | 178 |
| 90 | Thorium | Th | 4 | 2 | 1 | Radiactive matter | Light dark matter | 360 | 180 |
| 91 | Protactinium | Pa | 3 | 2 | 1 | Radiactive matter | Light dark matter | 273 | 182 |
| 92 | Uranium | U | 4 | 3 | 2 | Radiactive matter | Light dark matter | 368 | 276 |

**Figure 43.** Qperiodic table – 4 (Bent Rolf Pettersen).

| Number atom * | Name | | Plus quarks | Energy quarks | Other dim. | Category | Category 2 | Ilefos potential | Energy potential |
|---|---|---|---|---|---|---|---|---|---|
| 93 | Neptunium | Np | 4 | 2 | 1 | Radiactive matter | Light dark matter | 372 | 186 |
| 94 | Plutonium | Pu | 4 | 3 | 2 | Radiactive matter | Light dark matter | 376 | 282 |
| 95 | Americium | Am | 4 | 2 | 2 | Radiactive matter | Light dark matter | 380 | 190 |
| 96 | Curium | Cm | 4 | 2 | 2 | Radiactive matter | Light dark matter | 384 | 192 |
| 97 | Berkelium | Bk | 4 | 2 | 2 | Radiactive matter | Light dark matter | 388 | 194 |
| 98 | Californium | Cf | 4 | 2 | 2 | Radiactive matter | Light dark matter | 392 | 196 |
| 99 | Einsteinium | Es | 4 | 3 | 2 | Radiactive matter | Light dark matter | 396 | 297 |
| 100 | Fermium | Fm | 4 | 3 | 2 | Radiactive matter | Light dark matter | 400 | 300 |
| 101 | Mendeleviu m | Md | 4 | 3 | 2 | Radiactive matter | Light dark matter | 404 | 303 |
| 102 | Nobelium | No | 4 | 3 | 2 | Radiactive matter | Light dark matter | 408 | 306 |
| 103 | Lawrencium | Lr | 4 | 3 | 2 | Radiactive matter | Light dark matter | 412 | 309 |
| 104 | Rutherfordiu m | Rf | 4 | 3 | 2 | Radiactive matter | Light dark matter | 416 | 312 |
| 105 | Dubnium | Db | 4 | 3 | 2 | Radiactive matter | Light dark matter | 420 | 315 |
| 106 | Seaborgium | Sg | 4 | 3 | 2 | Radiactive matter | Light dark matter | 424 | 318 |
| 107 | Bohrium | Bh | 4 | 3 | 2 | Radiactive matter | Light dark matter | 428 | 321 |
| 108 | Hassium | Hs | 4 | 3 | 2 | Radiactive matter | Light dark matter | 432 | 324 |
| 109 | Meitnerium | Mt | 4 | 3 | 2 | Radiactive matter | Light dark matter | 436 | 327 |

* Number of atom element: Neutron, Proton

**Figure 44.** Qperiodic table – 5 (Bent Rolf Pettersen).

## Chapter 27

# Articles

I have written some posts and articles to better understand the relationship between the elements in the ilefos model. Here I also will try to show weaknesses in our present theories. I also suggest how we can improve these in a way which we can combine quantum mechanics with astrophysics.

> "Quantum mechanics and general relativity (which astrophysics is based upon), can't be combined. A correct physics should work in all fields. This indicates that current theories of physics might be wrong.
> General relativity is based on folding of a hypothetical fabric nobody can explain. Quantum mechanics also use hypothetical elements. When we use mathematical proofs without understanding what cause the observations, understand cause-effect, we often introduce hypothetical elements.
> If we do not work these theories, evolve them to find a clear cause-effect explanation which work in all fields, we will not have a correct physics. Without a correct physics we will not be able to correct explain nature. This might be the reason why we can't explain nature basics like gravity, dark energy (approx. 68% of the universe is dark energy), dark matter and the rapid accelerating expansion of the universe (a speed much higher than the speed of light)."
>
> Bent Rolf Pettersen

# Chapter 28

# Paradigm Shifts and Research

Gravity, dark energy, dark matter and the accelerating expansion of the universe – these are some of nature's greatest puzzles that scientists have so far been unable to explain. We can describe and calculate the phenomena, but why are we unable to explain them? It seems that the subject falls within what can be characterised as a paradigmatic debate. It is not always easy to be aware of when a subject is involved in such debates. It may therefore be helpful to look more closely at what characterises them.

> 'It is recognised that theories are only used because there are no better alternatives, not because they represent the final answer. One realises that it is not possible to come up with new solutions, and thus new knowledge, but that it is necessary.' (Professor Per Arne Bjorkum)

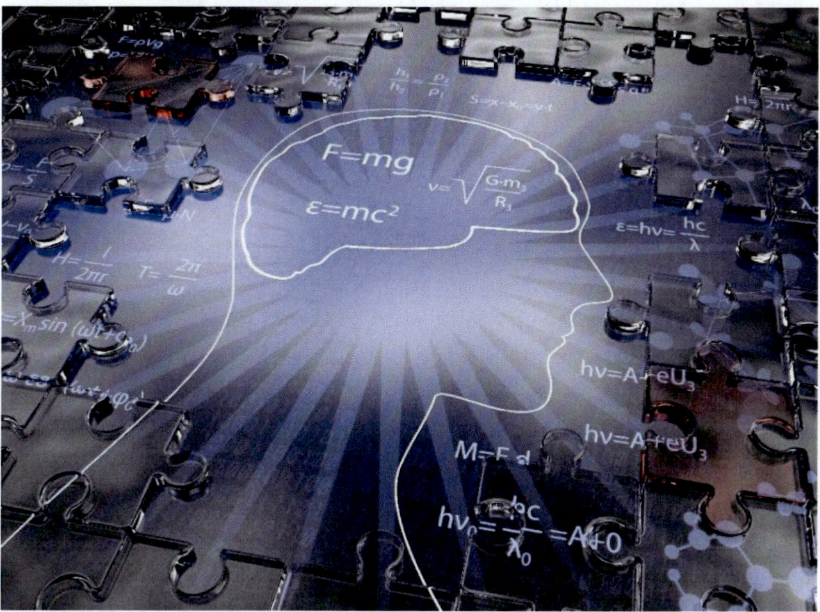


**Figure 45.** Brain (Istock).

Professor Thomas S. Kuhn (1922-1996) was a physicist and science historian who worked a lot with sociology of knowledge. He is most known for his works around paradigms. A paradigm is research shared and perceived as decisive within a research community.

A paradigm contains rules for good conduct which governs research in the specific field. The paradigm limits possible theoretical choices. The current paradigm will generally not be questioned. A paradigm is normal science that is not to be questioned.

Kuhn believed that if the current paradigm fails to provide fundamental answers, if many anomalies are detected, can this shake the paradigm. One can then try to find a new paradigm which can solve the anomalies. This can cause conflict with supporters of the old paradigm. If there is no common ground, major conflicts might arise between the groups. Are today's theories in physics examples of being within insoluble research paradigms?

Today's theories are based on a model of how the atom is put together that dates back more than 120 years. Some currently accepted theories about energies and gravity are close to 250 years old.

Natural science is largely presented in mathematical terms and formulas, so that they can be used to make calculations without having to think about cause and effect. If you want new knowledge about nature, you have to think about how something is connected – about cause-and-effect.

Even when we see that the observations do not match the theory, we are unwilling to depart from it until an alternative theory has been presented.

This is a paradox that can be an obstacle to innovation in research: Most scientists prefer scientific news that underpins what they already know and not news that challenges what they have built their career and self-image on.

When revolutionary news is launched, it can put the entire research community to the test, both intellectually and emotionally. When ideas are presented that go against the grain, they can bring about great and dramatic changes in theories about nature. Such dramatic changes can be called paradigm shifts (Thomas Kuhn).

A paradigm shift makes it necessary to build new theories almost from scratch. The process must start with new definitions and concepts, and new contexts must be established that turn established perceptions and theories on their head. Established scientists' views of the world can then feel threatened. Most scientists are guided by their theories and they are very attached to them. The theories often merge with the scientists themselves, and become part of their identity.

If a problem remains unresolved after decades of intense research in a field, it is reasonable to assume that the research community is not working on the right pieces. They are probably working within a paradigm that does not allow a solution.

If you are not aware of this possibility, you remain in this paradigm without the possibility of solving the task you have set out to solve.

The reason people get stuck in a paradigm is because it is common to perceive theories that are established and being practised within an established paradigm as proven. We are not trained to test the paradigm itself.

When a problem has existed for a while without anyone having been able to solve the task, it is usually the methodological approach that needs to be changed. You have to tackle the issue in a new way.

You will solve some, but not all, problems if you are true to a paradigm, or the established methodology. However, a greater and greater discrepancy between theory and observations will often accumulate.

Today, we have many different theories all based on what we want to be explained. In physics, there are different theories dealing with particle physics and astrophysics. These include the standard model, string theory, quantum mechanics and relativity. Each theory is often adapted to its area and needs to be changed when applied to other areas. There is no theory that can be applied to all areas.

Stephen Hawking spent much of his life trying to reconcile these theories into a universal theory that could be applied to all areas. He did not succeed. Is the reason that we find ourselves in a research paradigm which cannot give us the answers as Kuhm describes?

For such a theory to emerge, we probably need a paradigm shift in physics. We must dare to think outside established theories and models. Only then will we have a paradigm shift and a better understanding of nature.

# Chapter 29

# General Relativity, Is a Theory with Weaknesses – New Approaches to Gravity and the Expansion of the Universe

## Introduction

Isaac Newton quantified gravity in "Philosophiæ Naturalis Principia Mathematica" [1]. He described gravity and attraction between object as a force which originated from the objects.

In 1915 Albert Einstein published General Relativity (GR), where he combined gravity, space and time in one theory [2]. In 1919 the theory was confirmed in a solar eclipse, where the bending of starlight by the sun was more accurate than Isaac Newton's formula. GR's geometrical calculation of gravity is acknowledged as the theory of gravity on modern physics.

GR can't be conciliated with quantum mechanics. A correct theory of gravity should work on all fields. In this article I discuss elements which indicates weaknesses in GR. Can we trust GR to be a correct theory of gravity? I also suggest new approaches to explain gravity, dark energy, and the expansion of the universe.

## Discussion

Isaac Newton published his law of universal gravitation in "the Principia "(1687):

$$F = G\frac{m_1 m_2}{r^2}$$

**Figure 46.** Isaac Newton law of universal gravitation.

Here F is the gravitational force acting between two objects, m1 and m2 are the masses of the objects, r is the distance between their masses, and G is the gravitational constant.

This theory was suspended by Albert Einstein's General Relativity (GR). The gravitational constant was kept in GR.

Fabric of time and space is essential in general relativity. Gravity is calculated as a curvature of this fabric. This provides a precise calculation of gravity.

## Fabric of Space and Time

Gravity is in general relativity a geometrical calculation of the curvature of fabric of space and time. We have no explanation of this fabric. General relativity is based on folding of a fabric nobody can explain. This might indicate a weakness in the theory.

**Figure 47.** General relativity (iStock).

## The Accelerating Expansion of the Universe

The expansion of the universe is explained by general relativity as the expansion of the fabric of time and space. The fabric itself expands; it changes in scale.

Stephen Hawking pronounced in 2017 that he through observations could prove the general relativity wrong. If the fabric of space expanded, why did areas with matter not expand? Areas within galaxies and in matter seem to be constant, and do not expand. They seem to be constant mosaics in an expanding universe. If the fabric of space and time expands, all space should expand. When areas with matter do not seem to expand, this might disprove the theory of general relativity [3].

## Speed of Light

In 1905 Albert Einstein introduced the speed of light, c, as a constant, which also was essential in general relativity. The speed of light was the absolute speed, nothing could move faster [4].

The universe expands with an accelerating speed which is higher than the speed of light. This is explained by the expansion of fabric of time and space.

But is the speed of light the absolute speed? Light consists of photons, particles of light. These particles have energy and gravity properties. In gravitation lensing we see light being bent by strong gravity.

Around 68 percent of the universe is dark energy. This is estimated by gravity, or negative gravity, which influences galaxies in the universe. If dark energy influence matter, it should also influence photons. Dark energy is estimated to dominate "empty" space. Dark energy should therefore create a drag to photons. This drag might prevent photons to reach a speed higher than c.

If photons are affected by gravity, they should move a tiny bit faster towards strong gravity, and a tiny bit slower from strong gravity. This might indicate that the speed of light might have tiny variations through a doppler effect.

Pure energy units which are not affected by gravity, will not experience this drag of gravity/dark energy. Elements which are not affected by gravity therefore should not be limited by the speed of light, they should be able to move in speeds much faster than the speed of light. Pure energy units can be examples of elements moving faster than the speed of light.

In quantum entanglement we can synchronize spin of atoms/ objects. If we change spin of one particle, the other particle also change spin. This effect can be observed over long distances. The objects seem to communicate with a speed much faster than the speed of light. This is sometime explained by the state of dynamic involving the particle's wavefunction. The connected particles are communicating, either through an energy or wave function, which travel much faster than the speed of light.

This connection between synchronized particles, where the communication seems to be faster than the speed of light, can indicate speed of light is not the absolute speed. One of the foundations of general relativity, is the speed of light being the absolute speed. This again might indicate a weakness in general relativity.

## Observations

We have abundance of astronomical observation which seem to confirm general relativity. Still, we have some observations which seem to find it not accurate enough.

In Jody A. Geiger's paper "Jody A. Geiger Discrete Approach to Star Velocity Resolves Dark Matter Phenomenon" [5], he writes that star velocities near the galactic core in the Milky Way more closely match Newton's expression.

Even though most observations seem to be in accordance with general relativity, we have some examples where it does not seem to be precise.

General relativity makes good calculations of gravity, but it seems to not fit all observations.

## Dark Energy and the Expanding Universe

In matter we have an attractive and repulsive force. The attractive force creates attraction between atoms. The attraction between atoms creates atomic bindings and gravity. The attractive force close to atom's nuclei, where the force is strong, creates strong atomic bonds. The attractive force seems to decrease with distance from the nuclei, performing a weaker binding, gravity.

The repulsive force between atoms prevents nucleus from colliding with each other. This force seems to only work close to the nuclei.

Can the expanding universe be explained by these forces?

If we assume fabric of time and space do not expand, the universe itself has to move outward from its origin, Big Bang [6].

We first thought our universe was static. Then observation told us the universe is expanding. What can cause the universe to expand in an accelerating speed?

First, the force has to be stronger than the attractive force to matter (gravity) in the visible universe. In the term matter, I include dark matter, stars, galaxies and black holes. The gravity from these elements in the visible universe should work as a rubber band trying to pull the elements together again. We see matter in local areas, where galaxies seem to attract each other and merge. But the gravity between these local areas seems to not be strong enough to pull all matter in the universe together. A stronger force must explain the universe's expansion. Dark energy might be the explanation to this force.

Through the Lambda-CDM model [7], 68% of the visible universe is explained as dark energy (Λ). The measurement of cosmic microwave background seems to be indication of dark energy. Dark energy is explained as repulsive gravity (force), which may cause the expansion of the universe.

But this do not accurate explain why the universe expands in an accelerating speed. In order to explain this, I have to introduce new properties to dark energy.

Dark energy seems to be energy units which is not accumulated in quarks, matter and universal phenomena. Dark energy is then "free" energy units which is not connected to matter and universal phenomena.

These energy units are most likely the same energies we find in matter and universal phenomena. We can then assume they might have the same properties as we see in these elements. Dark energy then can be composed of different energy units which have attractive force, repulsive force and the other forces we see in matter and universal phenomena.

In matter, the attractive force seems to dominate the other forces at distance. The other forces seem to work at closer ranges.

If dark energy also is dominated of attractive energy units, dark energy, 68% of the visible universe, will create an attractive force. If we assume dark energy is somewhat evenly distributed, where can we find most dark energy? Does the space outside our visible universe also consist of dark energy? If so, this space must be approximately 100% dark energy. Most of the dark energy is then located outside our expanding universe.

In magnets the attractive force is stronger the closer we are to the source (magnet). If we assume this property can be transferred to the expanding universe, the attractive force from dark energy will be stronger the closer we come to strong dark energy.

If dark energy is mostly an attractive force, the attractive energy units may in calm areas, merge to fields of dense dark energy, which may have a very strong attractive force. These areas might surround our visible, expanding universe. The attractive force working, will then become stronger the closer we come to these areas. This can create an accelerating expansion of the universe. Dark energy can then be called the great attracter, which is responsible for the accelerating expansion of our universe. If this attractive force works at faster than the speed than light, this can explain why the visible universe expands at a speed faster than the speed of light.

## Gravitational Waves

In 1916 Albert Einstein predicted gravitational waves [8]. These were explained as ripples in the metric of spacetime which travel at the speed of light. In 2016 the team around LIGO detected gravitational waves from two black holes merging [9, 10].

Gravitational waves are sometimes described as a ring of freely floating particles released from a phenomenon which produce gravity. These particles affect the fabric of space by producing wave-like patterns.

An alternative explanation of gravitational waves:

Gravity might be gravity energy units, which attract each other and form gravity tracks. These free gravity energy units, which I call ilefos, may also form waves when traveling through space. If large amounts of ilefos energy units (gravity) travel in the same direction, they will clump together as waves of ilefos energy units due to their mutual attraction. These may be perceived as waves of gravity.

If the source of these gravity energy units, send out these units in variable strength, this might increase the gathering of gravity units, thus make a stronger wave pattern of the gravity energy units. When releasing the units in variable strength, I mean the number of gravity energy units released varies, not the strength of each energy unit.

These gravity energy units, which I call ilefos, attracts other units which have gravity tracks. A gravity track attracts other gravity tracks and particles which are affected by gravity. Photons are affected by gravity. We see this in gravitational lensing, where light is bent by strong gravity. LIGO observed a change in frequency in photons in lasers. Waves of ilefos will affect the photons, which should change the frequency as observed by LIGO.

Gravity waves are in this model, waves of gravity energy units traveling through space, and not distortion of the fabric of space.

## Conclusion

We can't explain the fabric of time and space. General relativity is geometric calculations of bending and distortion of this fabric. These calculations give us very accurate calculations of gravity, but they are based on a hypothetical fabric we can't explain. We have observations of stars close to strong gravity which seem to behave more according to Isaac Newton's calculations.

In general relativity the fabric of space itself expands. Yet, in areas with matter and stars, like galaxies, the space seems not to expand. They seem to be mosaics of non-expanding space in an expanding space. If space expands, it should expand in all areas, not leaving mosaics of non-expanding areas.

In general relativity light of space is the absolute speed. The universe expands with an accelerating speed faster than the speed of light. In quantum entanglement matter communicates with a speed faster than the speed of light. This might indicate that the speed of light is not the absolute speed.

These factors might disprove general relativity. General relativity seems to lack a cause-effect explanation.

An alternative explanation might be based on gravity energy units which attract each other. Due to their mutual attractive force, these units might form tracks of gravity units, gravity tracks. When different gravity tracks meet, the attractive force of the tracks create gravity. The force of these tracks decreases with distance from source much like Isaac Newton's calculations. Gravity tracks decrease in force due to the accelerating of the gravity energy units, creating longer distance between each unit. With increased distance between the units, the gravitational force will decrease with the length of the gravity track. Isaac Newton's calculations indicate this effect, but the calculations are not accurate enough and should be edited.

Dark energy seems to be the great attracter, causing the universe to expand in an accelerating speed. If dark energy is free energy units not connected to matter and universal phenomena, dark energy should consist of the same energies we find in these. The attractive force seems to be a dominating force working at distance in matter and universal phenomena. The energy creating this force should also give dark energy attractive properties. This attractive force can make dark energy accumulate into concentrated dark energy fields in calm areas outside our visible universe. These dark energy fields can create an attractive force pulling the universe, thus creating an expanding universe. This attractive force might be stronger the closer to the gravity fields we come, thus creating an accelerating expansion.

I call this model the ilefos model, based on the name of these gravity energy units, which I call ilefos.

## References

[1] Isaac Newton's Principia, the mathematical principles of natural philosophy (1846).
[2] Albert Einstein, "Die grundlage der allgemeinen relativitatstheorie" (1916).
[3] Stephen Hawking & Thomas Hertog, "A smooth exit from eternal inflation?," *Journal of High Energy Physics*, 2018.
[4] Albert Einstein, "On the electrodynamics of moving bodies," *Annalen der Physik*, 1905.
[5] Jody A. Geiger: "*Discrete Approach to Star Velocity Resolves Dark Matter Phenomenon,*" ResearchGate, 2021.
[6] Georges Lemaître, "*A Homogeneous Universe of Constant Mass and Growing Radius Accounting for the Radial Velocity of Extragalactic Nebulae,*" 1927.
[7] Michael S. TURNER, Lambda-CDM model, "ΛCDM, Much more than we expected, but now less than what we want," *Foundations of Physics*, 2018.
[8] Albert Einstein, "*On Gravitational Waves,*" 1918, Königlich Preußische Akademie der Wissenschaften.
[9] The LIGO Scientific Collaboration/the Virgo Collaboration, "Observation of Gravitational Waves from a Binary Black Hole Merger."
[10] Rainer Weiss, *"LIGO and the discovery of gravitational waves,"* 2018.

# Chapter 30

# Atomic Phase Displacement – Conversions of Energies in Atoms

## Abstract

We observe atoms emit several forces: An attractive force, a repulsive force and the strong nuclear force. The nature of these forces is not certain. We have different theories of how the forces occurs. We might assume each force originates from an energy. If so, the atoms quarks must store and handle several different energies. To perform the atom's tasks, each atom may require different energies. With abundance of energy, different types of atoms seem to prefer specific energies. Iron seems to convert electricity to magnetism. The energy conversion, Atomic Phase Displacement, seem to vary from each type of matter. In this article I discuss Atomic Phase Displacement, and I try to not link this to a specific theory. I focus on the phenomena Atomic Phase Displacement.

## Introduction

Atoms can convert energies. Different atoms seem to prefer different energies. Copper seems to prefer electricity. Iron seems to prefer magnetism. When atoms have surplus of energy, the atoms convert energies to the preferred energy. This is Atomic Phase Displacement, one of the key features of atoms.

## Energy Balance in Atoms

Atoms use a lot of energy to perform their main tasks. These tasks are sending out attractive force (performing atomic bindings and gravity), sending out repulsive force (preventing nuclei from colliding) and the strong nuclear force (holding the nucleus together). We also know atoms communicate with each other (quantum entanglement) and share energies. Constantly performing these tasks require a lot of energy. Atoms use energy to perform the atomic actions. To be in energy balance, the atoms must receive equal amount of energy.

Around 70 percent of our visible universe is dark energy. Dark energy might be described as free energy units, meaning energy units which is not organized in matter and universal phenomena.

Atoms must continuously receive the same amount of energy in which they use performing their core tasks. Most of the universe is dark energy. We therefore might assume they are constantly fed energy from dark energy to be in energy balance. They most likely receive these energies through dark energy harvesting tracks.

## Atomic Phase Displacement

Atoms seem receive different types of energies from dark energy. They also receive some energies from atoms in close proximity and free photons. Different types of atoms prefer different types of energies. The quark composition seems to decide which type of energy different atoms prefer. The periodic system lists matters and some properties due to their attractive force (gas, solid, metal).

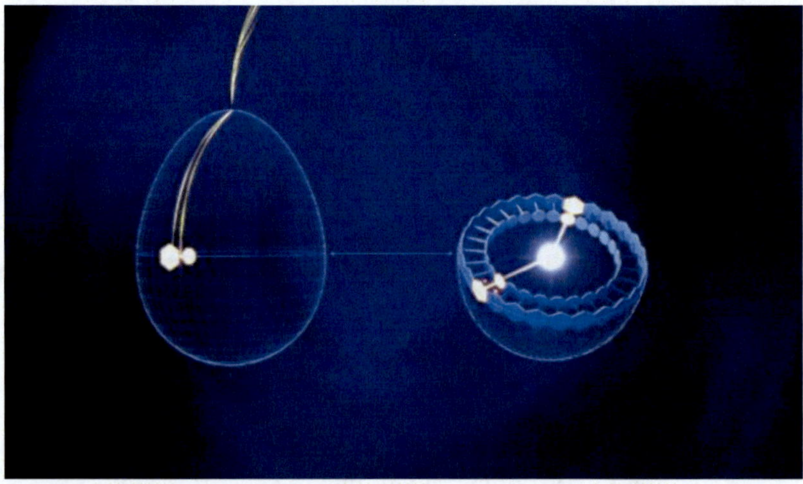

**Figure 48.** Atomic Phase Displacement in an atomic core element/proton. An atom converts an energy to another energy (Ola Tandstad).

I would like to focus on two main elements: The attractive force (atom bindings, gravity, magnetism), and dielectricity (electricity).

When we send strong electricity to iron, iron convert the electricity to an attractive force, magnetism. We see this in electromagnets. If we reverse the process, we have a generator. In a generator, copper convert magnetism to electricity. These conversions are examples of Atomic Phase Displacement.

Atoms also convert to other energies which they need to perform their other tasks, like the repulsive force, the strong nuclear force, communication energies, and other energies.

Atoms receive different types of energies from dark energy and neighbour atoms. Each atom converts these energies to the energies it needs to perform its main tasks.

## Conclusion

Atoms seem to demand different types of energies to perform the atomic actions. The atoms perform attractive force, repulsive force, strong nuclear force and seem to communicate with atoms close by. Different types of atoms seem to prefer different types of energies in Atomic Phase Displacement. The power of the bindings between atoms varies between different type of matter. The periodic system classifies these as gas, solid and metal. The attractive force emitted by different types of atoms vary. The amount of attractive force emitted, seem to be dependent on how much the atoms prefer this force in Atomic Phase Displacement. Atomic Phase Displacement is one of the key features of atoms, which I believe we should put more focus on. A more complete understanding of the processes in atoms will increase our understanding of nature.

# Chapter 31

# How Atoms Release Energy to Preserve Their Structure

## Introduction

$E = MC^2$, the mass-energy equivalence by Albert Einstein [1], tell us about the relationship between mass and energy. The law of conservation of energy states that the total energy of an isolated system remains constant. The energy can though be transferred to another form.
Atoms perform constantly several tasks: An attractive force, a repulsive force, the strong nuclear force and others. Do atoms loose energy when performing these tasks? If so, how can atoms be in energy balance? Atoms perform a repulsive force to prevent collisions between atomic nuclei. Such collisions can destroy atoms. Are atoms aware of their surroundings and can they adjust the repulsive force? How do atoms work to preserve their structure?

This article I discuss these matters and give explanations of these atomic actions through a new approach which explain conversions of energies in atoms, Atomic Phase Displacement. The new approach also shows how atoms can be in energy balance even though they constantly perform tasks in which they use energy. Through an organism metaphor, I also will show how atoms react to their surroundings and release energy to preserve their structure.

## Theories

This article will describe actions performed by atoms. I will not link these actions to specific theories. Our present theories in quantum mechanics have to use hypothetical elements in order to work. This may indicate weaknesses in the theories. I will therefore focus on the forces and actions performed by atom.

## Forces Performed by Atoms

Atoms constantly perform several forces. Atoms seem to be able to adjust these forces based on their surroundings. Atoms constantly perform these forces:

- *An attractive force*. The attractive force pulls atoms towards each other's. This force makes atoms connect and perform atomic bindings where the force is strong close to the atoms. This force seems to weaken with distance from the atoms. With distance from the atoms the reduced force seems to perform a weaker binding, gravity. Different type of atoms seems to have different level of the attractive force. Different levels of the attractive force might be caused by different quark composition in atoms.
- *A repulsive force*. The attractive force pulls atoms together. If atoms' nuclei collide, this can destroy the atoms' structure. The repulsive force makes atoms keep distance to each other and prevent collisions between the nuclei.
- *The strong nuclear force*. This force holds quarks together in special elements, protons and neutrons. Different quarks seem to concentrate and store different energies, which often is explained as different strings. The strong nuclear force seems to preserve the structure of quarks and elements of the nuclei.
- *Communication tasks*. The atoms seem to be able communicate with each other, thus be aware of their surroundings. In quantum entanglement we see how particles and matter can communicate and react to each other very fast over large distances. This interaction seems to be with a speed much faster than the speed of light. We can assume atoms communicates through different energies in close proximity and further away. This constantly communication make atoms aware of their surroundings.
- *Share energy through dielectricity*. Dielectricity seem to be a preferred energy to use when atoms share energy. This energy seems to be the easiest energy to convert. This energy might be shared as pure energy units through energy tracks or as energy stored in particles. I therefore call the energy dielectricity, which do not relate to a specific particle. I will discuss the conversion of energies later in the article.

Atoms seem to constantly produce these forces. Do the production of these tasks require the atoms to use energy? If so, they must constantly receive the same amount of energy to keep a constant energy level according to law of conservation of energy. How can atoms be in energy balance if they release energy to perform their tasks?

## Energy Balance in Atoms

If atoms constantly use energy to perform their tasks, atoms loose energy. If we assume the attractive force is energy units which is attracted to each other, they might form gravity tracks. These tracks are more dense and stronger close to the atoms' cores where they might form atom bindings. It seems like the tracks become less dense and weaker with distance from the atoms. This weaker attractive force is gravity. If this weakening of the gravity track continues according to Isaac Newton's formula for universal gravitation, $F = G\ (m_1\ m_2)/r^2$ [2], the track might become so weak that the attraction between the gravity units, might not be strong enough to hold the units together. Then the gravity track might dissolve into "free" energy units which are not connected to matter and universal phenomena. This is the definition of dark energy, "free" energy units which is not connected to organized energy (matter/ universal phenomena). Atoms then constantly produce dark energy.

To be in energy balance atoms must constantly receive energy. Where can we assume atoms can constantly harvest energy?

According to the Lambda -CDM model [3], 68% of the visible universe is dark energy ($\Lambda$). Most of the environment around atoms is dark energy. We can therefor assume that atoms constantly harvest dark energy to be in energy balance. The atoms must have special harvesting tracks to collect free energy units/ dark energy. The atoms then recycle energy units through dark energy to be in energy balance.

## Atomic Phase Displacement – Conversion of Energies in Atoms

Atoms do constantly perform several tasks which need specific energies. The demand for these tasks might vary due to the environment and the level of energies stored in specific quarks. To balance this, atoms must constantly convert energies. I call this Atomic Phase Displacement [4].

If atoms receive more energies than they can store in their quarks, they must pass on this energy. Dielectricity looks like the preferred energy for energy sharing between atoms. This energy seems to be the easiest energy to convert.

Atoms seem to prefer to share energy with dielectricity energy units. These units might be shared in dielectricity energy track or concentrated in special quarks which they release. We then must assume atoms have energy tracks connecting the atoms in the same way as gravity tracks. The atoms can adjust energy/share energy through these tracks.

If atoms receive abundance of energy, these tracks might not have capacity to transport the excess energy. The atoms then have to produce special energetic quarks in which they concentrate this energy and release these quarks. If we send high energy to matter, matter might produce heat and photons. Photons is particles (quarks) released by the atoms to get rid of surplus energy.

## Atoms Awareness of Surroundings

To know what energies the atoms need to produce, atoms need to be aware of their surroundings. To have even energy level, atoms must communicate with atoms in the vicinity. The atoms also have to communicate to know the location and other properties of atoms close by.

In quantum entanglement [5, 6, 7, 8, 9] we know matter and particles can communicate and coordinate spin/properties. This communication happens very fast, faster than the speed of light, and over great distances. We must assume atoms can communicate with special dimensional energies which is not affected by gravity and therefore travel much faster than the speed of light. If we assume atoms communicate with each other, atoms are aware and can adapt to their surroundings. This again might open up for a new approach to atoms. The atoms might be seen upon in an organism metaphor.

## The Release of Energies to Preserve the Structure of Atoms

If we assume atoms are aware of their surroundings and can convert energies through Atomic Phase Displacement, the atoms then must have a management system. This operating system controls the energies and functions of the atoms.

Atoms' main task must be to preserve their structure. To do this, they must prioritize the energy which help maintain the atomic structure. The strong nuclear force helps an atom to hold together the quarks and atomic elements. The repulsive force help protect the atomic nuclei from collisions which might destroy the atomic nuclei or part of it. The atoms then will prioritize the energies responsible for these forces.

If atoms move toward each other, atoms increase the repulsive force to keep distance and prevent collision between the atoms' nuclei.

This will cause increased repulsive force in the area. This abundance of repulsive force will result in a high level of repulsive energy units in the area, which increase the concentration of dark energy. Dark energy is free energy units which are not connected to atoms. These energy units will again be absorbed by atoms in the area, which receive a peak of repulsive energy to the atoms. This energy is more than the quarks responsible for the repulsive force can handle. The atoms convert this energy to the most common energy for sharing, dielectricity. The increased dielectricity is sent out from the atoms in dielectricity tracks and as increased production and release of energetic quarks. This will be perceived as heat and light. We see this in friction and collisions of matter.

If the force and speed of the collision is too high, the atoms might not be able to produce enough repulsive force to prevent a collision between the atomic nuclei. We then will have an interaction between the nuclei.

If the collision is not too strong, the collision might knock loose one or more atomic elements, like neutrons. If the atom has enough energy, the atom will produce new quarks and replace the missing element(s). If not, the atom might survive as an atom with fewer atomic core elements. The atom then will be converted to a new periodic element with fewer neutrons.

The atomic element knocked loose from the nucleus, for example a loose neutron, will then be a free neutron which is not connected to an atom. This atomic element will preserve its structure for a short while. But without the atomic operating system and the strong nuclear force, the element will not be able to hold the quarks together in a structure. The element will then dissolve into free quarks. These might again dissolve and release their energy.

An atom which loses an atomic element in a collision and replace the element, will rebuild and maintain the atomic structure, but the atom will have reduced energy level. In time the atom will get back its energy through energy input from other atoms, loose energy units/ dark energy, and external quarks/ photons.

If the collision is between the nuclei are too strong, too many atomic elements will be knocked out from the nucleus. The nucleus will not have enough energy to reproduce the missing elements, and the strong nuclear force will not be able to hold the rest of the atomic core together. We then will have a total atomic dissolution. All atomic elements will be released and dissolve. This will result in an immense release of energy. We see this in a nuclear explosion.

These atomic actions we see in friction, fission and atomic dissolutions.

## Conclusion

Atoms perform several forces. The repulsive force prevent contact between nuclei. Atoms also can convert energies through Atomic Phase Displacement. In this process, atoms constantly convert energies what they need to perform their tasks and preserve their structure.

Atoms seem to be aware of their surroundings. If they sense a chance for collision which will harm their structure, they increase the repulsive force. This help atoms keep distance and prevent nuclei from colliding. Friction is an example of this. The increased repulsive energy can be perceived as heat.

Atoms can release energy through increased repulsive force to preserve their structure.

## References

[1] Albert Einstein, "Uber einen die Erzeugung und Verwandlung des Lichtes betreffenden heuristichen Gesichtspunkt," *Annalen der Physik*, 1905.

[2] Isaac Newton, "The Pricipia," 1687.

[3] Michael S. TURNER, Lambda-CDM model, "ΛCDM, Much more than we expected, but now less than what we want," *Foundations of Physics*, 2018.

[4] Bent Rolf Pettersen, "Atomic Phase Displacement – conversions of energies in atoms," *ResearchGate,* 2023.

[5] E. Schrodinger, "Discussion of Probability Relations between Separated Systems", *Mathematical Proceedings of Cambridge Philosophical Society*, 1936.

[6] Manoj Kumar Joshi, Christian Kokail, Rick van Bijnen. "Exploring Large-Scale Entanglement in Quantum Simulation," *ResearchGate,* 2023.

[7] Andreas Wendl, Heike Eisenlohr, Felix Rucker, Christopher Duvinage, Markus Kleinhans, Matthias Vojta, Christian Pfleiderer, "Emergence of mesoscale quantum phase transitions in a ferromagnet," *Nature,* 2022.

[8] "Quantum materials: Entanglement of many atoms discovered," *Science Daily,* 2022.

[9] Jia Kong, Ricardo Jimenez-Martinez, Charikleia Troullinou, Vito Giovanni Lucivero, Geza Toth, Morgan W. Michell, "Measurement-induced, spatially-extended entanglement in a hot, strongly-interacting atomic system," *Nature Communications*, 2020.

# Chapter 32

# Conclusion

Albert Einstein described the relationship between mass and energy in his mass-energy equivalence formula, $E=mc^2$ (Annus Mirabilis papers, Annalend der Physik, 1905). This formula might be audited, but it shows the relationship between energy and matter.

Energy can be organized in matter. Matter can be dissolved and release energy.

Over decades we have tried to find answers to the fundamental questions of nature without success. What is gravity? What is dark energy? What is dark matter? Why does the universe expand in an accelerating speed faster than the speed of light?

A correct description of gravity must work in both quantum mechanics and astrophysics. General relativity only works in astrophysics and uses hypothetical elements which nobody can explain. A correct physics should work in all areas and have a clear cause-effect explanation.

Are we limited within scientific frames which do not allow answers to these questions?

Professor Thomas Khun referred to these issues as research paradigms (Kuhn, T.S. 1962, The structure of scientific revolutions). In a paradigm shift we have to build new theories from scratch. One has to build new definitions and terms, and establish new relationships which turns established beliefs and theories upside down.

New observations can start paradigmatic debates. If a problem remains unsolved after decades of intense research in a field, is it reasonable to assume that the research community is not working with the correct pieces. You are then probably working within a paradigm that does not enable a solution. Absence of solutions should also start paradigmatic debates.

This book is based on the ilefos model. The ilefos model is an alternative conceptual model of energies and matter. In the model, I try to not lean on other theories which might have weaknesses.

Some basics with the ilefos model are:

- You need energy to grow matter. Strong energy creates matter in the universe.
- In fission and atom dissolution we release energy from matter. We see this in stars and black holes.
- Energy can be free energy units, as in dark energy, or energy organized in quarks, quark groups and matter.
- Energies can be material energies which dominates in matter, or dimensional energies which are not influenced by gravity or matter. Dimensional energies carry and organize information. These have touching points in matter.
- Different energies have different properties. The reaction between similar or different energies creates forces.

The physics of the ilefos model also explain the properties of atoms, the actions of atoms and the operating system of atoms. Through observations of quantum entanglement we know atoms can communicate with each other. Atoms seem to be aware of their surroundings and act upon feedback. They therefore seem to have an operating system and can be looked upon in an organism metaphor.

I hope the model can create a better understanding of nature and the dynamics of energies, especially around atoms and the behaviour of atoms.

# Index

## A



## D



## F



## I



## L



## N



## P



## Q



## R



## S

## T



## U